SpringerBriefs in Physiology

More information about this series at http://www.springer.com/series/10229

Kin Kheong Mah · Hwee Ming Cheng

Learning and Teaching Tools for Basic and Clinical Respiratory Physiology

 Springer

Kin Kheong Mah
Royal College of Medicine Perak
Universiti Kuala Lumpur
Perak
Malaysia

Hwee Ming Cheng
Faculty of Medicine
University of Malaya
Kuala Lumpur
Malaysia

ISSN 2192-9866   ISSN 2192-9874   (electronic)
SpringerBriefs in Physiology
ISBN 978-3-319-20525-0   ISBN 978-3-319-20526-7   (eBook)
DOI 10.1007/978-3-319-20526-7

Library of Congress Control Number: 2015942554

Springer Cham Heidelberg New York Dordrecht London

Printed on acid-free paper

Springer International Publishing AG Switzerland is part of Springer Science+Business Media
(www.springer.com)

καὶ ᾠκοδόμησεν κύριος ὁ θεὸς τὴν
πλευράν ἣν ἔλαβεν ἀπὸ τοῦ Αδαμ εἰς
γυναῖκα καὶ ἤγαγεν αὐτὴν πρὸς τὸν
Αδαμ

Septuagint (LXX) Genesis 2:22

*Then the LORD God made a woman from
the **pleura** he had taken out of the man and
brought her to the man*

Preface

This Clinical Respiratory Physiology book is written by two teaching colleagues, both of us with a special interest in communicating physiology to our students. The first section of the book will review the framework and foundations of basic respiratory physiology. Since this book was not written to be a comprehensive physiology text, we have focused on leading the students to appreciate and understand integrative principles and homeostatic mechanisms in lung functions.

Our purpose is to help synthesize and summarize pathways in respiratory mechanics, dynamics of air–blood and blood–cellular gas exchanges. We both strive to teach to make physiology fun to learn. This emotional aspect of knowledge acquisition is reflected in the way the topics are approached, e.g. by using playing cards in what is coined as 'Respi-CARDology'.

We breathe to live, we live to learn, and we hope that this book will stimulate both students/educators to discover the diverse ways to learn and enjoy physiology. We all can think and teach 'outside the (thoracic) box' in appreciating our lungs and how these doublet 'wind instruments' in our body are interlinked with other organ functions. For example, the lungs and the kidneys cooperate in maintaining ECF/blood pH. Thus, the lungs are an integral part of the physiologically orchestrated, symphonic homeostatic harmony in our body.

The second part of this book deals mainly with the application of the foundations of respiratory physiology. We are convinced that when students are able to successfully apply what they have 'learned' to clinical and non-clinical scenarios with respect to respiratory physiology, they have then reached another 'atmospheric' level in their understanding of respiratory physiology.

We hope that students will find this book useful to help them in appreciating the 'whys' and 'hows' of physiological processes in the human body and not resorting to memorizing facts and data merely to pass examinations.

Kin Kheong Mah
Hwee Ming Cheng

Contents

Part I
Respiration Essentials Revisited

Teaching Tools for Basic Respiratory Physiology

This respiratory Physiological learning title hopes to provide some useful handles to students to grasp the main essentials of lung breathing and its associated roles with the pulmonary and systemic blood circulation to ensure the optimal oxygen supply to cells. The blood and air transport processes for carbon dioxide from cells to lungs are also covered.

Comprehensive physiological texts are easily accessible. This learning supplement approaches respiratory system from various angles and perspectives to give students more foundational and conceptual appreciation and hopefully also enjoy understanding respiratory physiology.

Some of the teaching tools that have been used by one of the co-authors (HMC) among the students in the University of Malaya are shared here in this book. These include the use of music, drawings and recreational games.

Trumpet and Respiratory System

The eight notes of a scale ...*do re me* to the high *do*.

A trumpet is a wind brass instrument and is played with the aid of our respiratory system.

Let me trumpet out the 8 steps in respi-cardiovascular physiology that are essential in appreciating the pathways the two respiratory gases, O_2 ad CO_2 take between the external air and the cells in the body.

1. How air enters the lungs and moves out of the lungs in what we called the mechanics of respiration.
2. How diffusion of oxygen at the alveolar capillary membrane occurs.
3. How arterial blood transports oxygen to the cells.
4. How hemoglobin unloads oxygen to the cells.
5. How metabolic CO_2 is taken up by the capillary blood.
6. How venous blood transports CO_2 to the lungs.
7. How diffusion of CO_2 occurs at the alveolar capillary membrane.
8. How respiration is controlled to maintain the alveolar air CO_2 and O_2.

Balloon, Music Scores, Telephone, Playground, Playing Cards, Songs/Rhymes

Balloon: The use of the balloon helps to demonstrate a few major aspects of the elastic structures of the lungs and the chest wall, both acting as a functional unit. Physiological definition of elasticity is described as an opposing force to distention and stretching. The blowing of a balloon using an expiratory effort quite clearly illustrates this. The capacity to accommodate air or the compliance is thus inversely related to the physiological elastance.

Music Scores: Following on from the trumpet, the major musical keys of C, D, F and G are used to highlight several keywords in respiratory physiology. C is for the content of both oxygen and CO_2 in arterial and venous blood. D is for a normal 'Dead' space and also includes pathophysiological alveolar dead space in certain lung conditions. Diffusion of both oxygen and CO_2 is crucial for gas exchanges at the lungs and the tissues. The key of F denotes Fick's law of diffusion that considers several parameters that determine simple diffusion of both respiratory gases. Gravity is signified by G. Both alveolar ventilation and pulmonary circulation vary vertically in the upright lungs as a result of gravitational forces on intra-pleural pressure and blood hydrostatic pressure, respectively. Thus, the effect of gravity explains the changing V/Q matching from the base to the apex of the upright lungs.

Telephone: The pathway for both oxygen and CO_2 can be traced in a clockwise direction from lungs to cells and back to lungs again. At various points along this circuit, the partial pressures of O_2 and CO_2 are relatively stable and have universal values in all normal persons. These PO_2 and PCO_2 values in mm Hg along the lungs–cells–lungs journey are thus quite essential to keep in mind. By using the classical clockwise dial telephone and giving telephone numbers representing PO_2 or PCO_2, these serve as memory aids for handling blood gas data in clinical situation.

Playground: The three main play structures in any park are the slide, the swing and the see-saw. The slide is used to define the gradients that are so essential for oxygen and CO_2 diffusion. For example, one main purpose for the control of respiration is to keep constant the alveolar air PO_2/PCO_2. This will then ensure that gas exchange down the respective O_2/CO_2 partial pressure gradients at the alveolar-capillary membrane is adequate. The swing is a repetitive device, and this is similar to the respiratory cycle of inspiration and expiration. The see-saw is often balanced during play by many children on each side. As in other systems in physiology, respiration also functions homeostatically to maintain the correct balance of blood oxygen and blood CO_2 content. Blood pH balance is also a crucial function of breathing via the regulation of blood PCO_2.

Cartoons: Well-drawn cartoons are engaging and can include precise physiological mechanisms. Two examples are the O_2 unloading effect of Bohr (Sect. 2.8 'Bloody Multitasker Red Cell') and the primary hypercapnic control of normal respiration (Cover of Chap. 3).

Playing Cards: The clubs resemble the pulmonary alveoli. Respiratory physiology does include a number of values to take note of. Despite the creative use of playing cards, the clubs can be designed and effectively used to get use to certain quantitative aspects of lung physiology in small-group tutorials.

Songs/Rhymes: The time-tested nursery rhymes and children's songs stay with us. Composing of respiratory physiological songs by the students can enhance learning as the basic knowledge of the physiology has to be understood before any meaningful rhyme or song can be attempted.

Chapter 1
Mechanics of Respiration

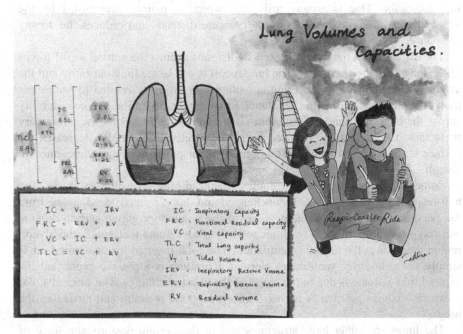

Credit Respiratory Roller Coaster by Adlina Athilah Abdullah, 3rd Clinical Yr, Universiti Malaya Medical School

Abstract In this chapter, we will look at the process of air entry and air exit from the lungs, namely inspiration and expiration respectively. The key explanatory words in describing breathing are pressures, forces, spaces. The pressures in turn include atmospheric, alveolar, intra-pleural, transmural/transpulmonary.

Keywords Lung recoil · Dead space · Effects of gravity · Trans-pulmonary pressure · Lung volumes

1.1 Introduction: Ready to Inspire and Breathe Air

Inspiration is an active process and is described as 'negative-pressure' breathing. The 'negative' here refers to the intra-alveolar pressure decreasing below the reference atmospheric pressure (0 mm Hg). The pressure gradient moves air into the lungs. In normals, the active contraction of inspiratory muscles makes the intra-pleural pressure more negative which leads eventually to the 'negative' intra-alveolar pressure that accounts for the inspired tidal volume of air.

Active inspiration requires energy to be expended. The active work of breathing is needed to overcome several opposing forces or resistances to airflow during inspiration. One is airway resistance which is mainly determined by the airway radius. The airway trans-mural pressure distends and reduces the airway resistance.

The expansion of the lung alveoli also distends against the natural recoil forces of the lungs and the chest wall. The lung recoil is due to the lung elasticity and the alveoli surface tension force. Lung compliance is inversely related to lung elasticity, the latter defined as the resistance to stretch or distention. Alveoli stability across the huge thousands of different sizes alveoli is effected by the pulmonary surfactant. The surfactant produced by alveolar cells differentially decreases the surface tension force to equalize the intra-alveolar pressure in the interconnected alveoli.

Some of the lung spaces or volumes change during a respiratory cycle. The non-respiratory space is basically unchanged and this is termed 'dead' space from the perspective of a non-exchange area for the respiratory gases, carbon dioxide and oxygen. The portion of fresh air in each tidal volume that enters the alveoli region determines the alveolar ventilation. The total inspired air or expired air is simply the pulmonary ventilation. Normal expiration is a passive event and the expired tidal volume is due to the recoil of the chest wall/lungs. The intra-alveolar pressure becomes positive or higher than atmospheric pressure and air leaves the lungs.

The lungs are rather large structures and in the upright posture, the force of gravity causes regional difference in alveolar ventilation. The effect of gravity results in a progressively less negative intra-pleural pressure from the apex to the base of the upright lungs. This leads to a difference in the trans-pulmonary pressure across the wall of the alveoli in the vertical axis of the lungs. The different sizes alveoli at the end of a normal expiration (or at the lung volume called functional residual capacity, FRC) do not have the same compliance. The alveoli at FRC towards the base of the upright lungs are more compliant and accounts for them to be better ventilated with the next inspired tidal volume.

1.1.1 Recoil Forces, Push and Pull

A good teaching illustration of recoil forces in respiratory mechanics is to get students to think about changes in the functional residual capacity (FRC) from the supine to the upright posture. A useful question to start the discussion is "How does the Inspiratory capacity compare between the supine and standing position?" Most students would say that the standing person has a greater inspiratory capacity and this natural response seems to stem from the quite logical prediction that the diaphragm will be less compressed in the upright person.

The students can be told that unless one is quite overweight, there is no significant compression effect by the abdominal contents on the diaphragm to affect breathing. Moreover, be definition, an Inspiratory capacity is a measured during a maximal inspiratory effort and this will annul any little increase in mechanical compression on the diaphragm.

The students are always surprised when they are told that in reality the Inspiratory Capacity (IC) is higher in the supine position. Again, remind them that the IC when performed takes the lung volume to its maximum total lung capacity.

Total lung capacity is the sum of the IC and the FRC. By deduction, the FRC must then be less in the supine position and a maximum inspiration will then account for the higher IC when lying down.

The next obvious question will lead to the consideration of why the FRC is lower when lying down compared to the standing posture. This is understood when the recoil forces of the lungs and the chest wall are observed. At the end of a normal expiration, the volume of FRC is determined by the opposite lung inward recoil and the chest outward recoil forces.

The next point for the students to note is that the mechanics of the chest wall and the diaphragm operates as a functional single unit. If the FRC is less lying down, this must mean that the chest wall recoil is reduced and the net recoil balance point will shift to a decreased FRC value. The decreased chest wall outward recoil is in this case, due to the slight abdominal compression on the diaphragm in the supine position.

By this to and fro, push and pull questions and feedback interactions, the Dr. Doolittle Physiology educator can lead the students to think, derive and understand physiological mechanisms for themselves.

1.1.1.1 Blowing Balloons and Inspiring Respiratory Knowledge

One effective self-directed learning exercise during Respiratory physiology classes is to distribute balloons for students to blow. The balloon is elastic and inflating the balloon slowly can be an impressive tool to teach a few aspects of respiratory mechanics. There are two significant teaching points here.

(a) Elasticity or Elastance is inversely related to the Compliance of the structure. Physiologic Compliance is defined as the change in the volume of the elastic structure per unit distending or expanding pressure. In physiologic structures, the elastic recoil opposes stretch. Thus the more elastic structures have a lower compliance. This is similar to the vascular structures where the elastic arteries are less compliant than the venous capacitance vessels.
(b) By slowly blowing up the balloon, it is the common experience that the beginning of the inflation is harder and further inflation towards the end of the near maximally expanded balloon also requires a greater expiratory effort. This demonstrates a similar compliance profile of the alveoli at different lung volumes. The graphical relationship between distending trans-pulmonary pressure (x-axis) and lung volume (y-axis) is a sigmoid shaped line. This shows that the lung compliance also changes and varies with lung volumes. At low lung volume near residual volume and at high lung volume close to total lung capacity, lung compliance is lower compared with the lung compliance around the functional residual capacity (the steep portion of the lung compliance curve).

1.1.2 PhyBalloon Learning (PBL!)

Distribute some balloons during a tutorial class. Get settled down and warm up by inflating the balloons.

How can we use the balloon to teach some aspects of respiratory physiology? Firstly, the physiological definition of elasticity. **Elasticity** is not defined as the stretchability of a physiologic structure unlike the common conversational usage when we are shopping for an elastic trousers. Rather, an elastic structure will recoil when distended and a recoil force is produced. This recoil force opposes stretch. Thus it takes more effort to expand a highly elastic structure than one with minimum elasticity.

We also use the term **'Compliance'** as a parameter to reflect the distensibility of a structure. Compliance is inversely related to the elasticity. Compliance is the volume change of a structure when a unit pressure is applied to stretch it (dV/dP). Thus the more elastic a structure, the less compliant it will be e.g. veins are less elastic than arteries but veins are more compliant and function as capacitance vessels ('blood reservoir').

Ask the students to blow again the balloon and then pose the question "Is the compliance of the balloon the same during the inflation from a small to a near maximum volume?" It will be experienced during the blow-up that a greater effort is needed to expand the balloon at the initial stage when the volume of the balloon is small. Also as the balloon reached its maximal size, further pressure of blowing air also expands the balloon to a smaller degree.

This can be used to compare the changing lung compliance during lung volume changes from residual volume (RV) to total lung capacity (TLC). The shape of the lung compliance curve (dV on y-axis and dP along x-axis) is sigmoid. This reflects the analogous situation in the balloon i.e.teh lung compliance is less nearer RV and approaching TLC. The better lung compliance is at the lung volume changes around the functional residual capacity, the lung volume after a normal expiration.

In the lungs, the Recoil force is partly the Elastic recoil and the distending trans-pulmonary pressure to expand is also opposed by the recoil due to the Alveolar Surface Tension.

1.1.3 An Alveolus Is 'Laplace' to Be!

The community or population of alveoli are networked with each other via the same air they are surrounded by. Size wise they are not homogenous. In the upright lungs, at the point of normal expiration (FRC) before the next breath is taken, the apical alveoli are slightly more expanded than the basal alveoli. This size variation is due to gravitational effects on intra-pleural pressure and thus the trans-mural pressure(trans-pulmonary pressure, Ppt). The apical alveoli are exposed to a higher Ppt at FRC. Even along the same horizontal level from the lung base, there can also be variability in the distribution of airflow, probably due to random differences in the diameters of bronchi formed at the bifurcation of airways.[1]

---

[1]Davies and Moore, The Respiratory System

The size heterogeneity presents a potential mechanical problem for all the alveoli to co-exist. This is explained by Laplace's law concerning the distending or expanding pressure that is required to keep an alveolus open. The other factors that determine the intra-alveolus pressure, P needed for the alveolus to remain open are the alveolar surface tension and the alveolar radius, given by $P = 2T/r$. I ask the students "Is the distending pressure required to maintain an open alveolus less or more in a smaller alveolus, assuming the surface tension is the same in all alveoli?" So Laplace's equation indicates that the holding pressure is higher for a smaller alveolus.

Since the alveoli are air-linked, it is not possible to have this potential heterogeneity in alveolar distending pressure. Theoretically, the smaller alveoli will collapse when its higher intra-alveolar pressure drives air into the larger alveoli where the alveolar air pressure is lower.

I then follow on with the next question "How do you think the lungs can compensate to have a stable population of heterogenous alveoli of different sizes?" Looking at Laplace's formula, the solution would be to alter the surface tension (ST) so that the ST of smaller alveoli would be reduced.

The reduction of ST would then make or equalize the ST/r ratio in all the alveoli and alveolar stability is achieved.

The students should now see clearly that this ST-lowering in the lungs is accomplished by pulmonary surfactant. The students should note that pulmonary surfactant is present in all alveoli. Thus the effect of surfactant on reducing ST must be greater in the smaller alveoli compared with the larger alveolar sub-population.

1.1.3.1 Respiratory Rhythms Are the Music of Life (Key of D)

The respiratory space from the mouth to the alveoli are divided into have two broad functions. The alveolar respiratory zone is the place of gas exchange. The conducting airway zone does not participate in gas exchange ('Dead' in this aspect) but serve as conduits for inspired and expired air. The conducting zone also provides the protective distance since the lung alveolar tissues are in direct contact with the external environment. The conducting airways are also lined with mucus and cilia as 'gatekeepers 'for the respiratory zone besides the presence of the mucosal immune system in the walls of the airways. The Key of D is this 'Dead' space, a misnomer since the space serves obvious essential physiologic roles.

Alveolar ventilation is calculated from considering the portion of the tidal volume that enters the respiratory zone and multiply this by the frequency of breathing. The normal anatomy of the branching respiratory tree gives labels the conducting airways as anatomical Dead space.

An alveolar Dead space is not normal and represents alveoli devoid or deprived of its blood perfusion, thus these affected alveoli become functionless as respiratory exchange areas. The ventilation/perfusion ratio of an alveolar dead space is then infinity. The compensation for an alveolar dead space is to divert the airflow to better-perfused alveoli via bronchoconstriction by regional hypocapnia in the affected alveoli. As in music notations where both the line and the space notes make their important contribution to the melody, the key concepts of respiratory Dead space (anatomical, alveolar) need to be understood in the normal and pathophysiologic lungs.

1.1.3.2 Respiratory Rhythms Are the Music of Life (Key of G)

G for Gravity. The effect of gravitational force plays a significant role in pulmonary physiology. Since the lungs are relatively large structures, gravity in the upright position affects both the pulmonary blood distribution and the alveolar ventilation at different vertical levels of the lungs.

Gravity increases the blood hydrostatic pressure progressively below the heart level and conversely, the blood pressure decreases with greater heights above the cardiac line.

Gravity also results in different degrees of negativity in the intra-pleural pressure in the upright lungs. The mass of the lung tissues produce a less negative intra-pleural pressure at the base of the lungs compared to the lung apices. Consequently, this leads to a better alveolar ventilation at the basal regions of the lungs.

The effect of Gravity thus affects both the alveolar ventilation as well as the pulmonary blood perfusion as one ascends up the upright lungs. The upright lungs is classically divided into three regions based on the ventilaton/perfusion changes produced by Gravitational force. We could musically code this as an ascending three octaves and each lung region ends as with an octave at the G note!

Gravity in the upright posture also increases venous pooling. The pulmonary blood volume is a significant portion of total blood volume. In the supine position, the thoracic pulmonary blood volume is larger than in the standing person.

1.1.4 The D Space Note

Yes the Dead Space is Lung Physiology. The Dead Space is actually a living entity in respiration. The meaning of the adjective 'dead' is narrowly employed in relation to non-gas exchange regions of the respiratory tree. Thus the anatomical dead space is simply the conducting airways.

Bear in mind that the lungs are in direct first exposure to the external environment. As such the multiple bifurcating and branching of the respiratory tree can be viewed as representing a protective labyrinth to minimize harmful, injurious agents from accessing into the depths of the lungs. We know that the conducting airways are not just passive conduits. Lining the airways are protective mucus and rhythmic ciliary motions that serve to push out unwanted airborne 'invaders'. In addition, the mucosal immune system also actively participates in healthy lungs.

In the conducting airways, the smooth muscle function are also affected by autonomic parasympathetic and sympathetic activities. Normal bronchial smooth muscle tone provide the main factor in normal airway resistance as airway resistance is inversely related to the 4th power of the radius. Hyper-responsive airway is one of the hallmarks of increased airway resistance in asthma.

At the end of a normal expiration, the lung volume, termed functional residual capacity (FRC) included the dead space volume of air in the conducting airways. The FRC can be considered as a buffer air zone that helps to prevent major fluctuations in alveolar air partial pressures of CO_2 and O_2. In chronic obstructive lung disease the phenomenon of 'airtrapping' overexpands the FRC with resultant decreased alveolar PO_2 and increased alveolar PCO_2.

So let's listen and hear the living music of the D Space note!

There is occasionally a jarring D space note. This is not the normal anatomical D space. It is called an Alveolar D space. The concept is the same, local changes in the alveolar region leads to no exchange of gases. Specifically, it is caused by any reduction or absence of blood supply to a part of the lungs. Those regional alveoli that are deprived of blood perfusion are no longer in exchange mode and become functionally 'alveolar dead space'.

1.1.5 The Key of G, the Key Effect of Gravity on Respiratory Physiology

This effect of gravitational force is most evident in the upright lungs. The sequential effects on alveolar ventilation, on pulmonary blood perfusion and also total thoracic blood volume are as follows:

1. Intra-pleural pressure is less negative at the base of the lungs due to the effect of gravity o the large mass of the upright lungs.

2. This leads to a larger trans-mural pressure at the alveolar wall (also termed trans-pulmonary pressure) of the apical alveoli, at the end of normal expiration (at FRC, alveolar pressure is 0 mm Hg).
3. The smaller basal alveoli has a higher compliance and at the next inspired tidal volume, more of the fresh air ventilates the basal alveoli.
4. The G factor also explains the difference in pulmonary blood flow at various heights of the vertical lungs. Gravity affects the hydrostatic fluid pressure of the blood.
5. Both the pulmonary arterial and pulmonary venous pressure is affected by gravity. The perfusion driving pressure is not altered as such.
6. The pulmonary blood flow decreases from the base to the apex of the upright lungs due to the additional alveolar air pressure that exerts an external compression on the blood vessels.
7. The G factor also accounts for the difference in total pulmonary or thoracic blood volume between the supine and upright person. Venous pooling when standing significantly decreases the intra-thoracic blood volume.
8. The functional residual capacity (FRC) is also less in the supine position due to Gravity and its influence on the abdominal content compression on the diaphragm-chest wall mechanics. Thus Inspiratory Capacity, the maximal inspired volume from FRC to total lung capacity is actually more in the supine position.

1.1.6 The Trumpet and Respiration

This brass blowing musical instrument can be an instructive demo tool in learning some key aspects of lung function.

All notes produced by a trumpet are achieved by a forced expiration!

During normal breathing, expiration is a passive event and achieved by passive recoil of the lungs and chest wall. Although at FRC, the chest wall recoil reverses and is counterbalanced by the lung inward recoil.

In a normal respiratory cycle, the intra-pleural pressure is negative throughout inspiration and expiration. The intra-pleural pressure (Pip) becomes more negative during the active phase of inspiration when the inspiratory muscles contract.

A forced expiration during trumpet makes the Pip positive. Coughing and sneezing also produce a positive Pip.

Even in normal expiration, the airway resistance is slightly higher than during inspiration but this obviously is academic and insignificant. The distending trans-mural pressure at the airways affect the airway resistance. The air passage trans-mural pressure (Ptp) is the difference between the inside and outside of the wall of the airway. The less negative Pip during expiration means the Ptp is also lower during expiration.

The students can be asked "How do you expect the alveolar air PO_2 and PCO_2 to change during the production of a prolonged trumpet note?" The normal partial pressures of CO_2 and O_2 are due to the equilibrium between the exchange diffusion of the respiratory gases at the alveolar-capillary membrane and the fresh input of alveolar ventilation. With a prolonged expiratory musical note, there is a prolonged period of no alveolar ventilation. Therefore the alveolar air PCO_2 will rise above 40 mm Hg and the PO_2 will decrease to below 100 mm Hg.

What about the pulmonary blood flow? is that affected with a prolonged trumpet note? Indeed, the trumpeter is functionally increasing her intra-thoracic pressure. This means the central venous or right atrial pressure will be raised and the venous return is momentarily reduced. Consequently, the right heart cardiac output, representing the pulmonary blood flow is decreased...Tadaaaaaaaaaaaaaaaaaaa.

Chapter 2
Diffusion, Blood O_2, CO_2 Content and Transport

Abstract This chapter describes how oxygen is transported from the atmosphere to the cells and also the transport of carbon dioxide from the cells to the atmosphere. The systemic and pulmonary vascular networks are designed to ensure optimal cellular respiration as well as removal of the carbon dioxide produced.

Keywords Ventilate · Oxygenate · Pulmonary circulation · Fick's law of diffusion · Blood gas content · Ventilation/Perfusion matching

2.1 Introduction: Ready to Oxygenate, Deliver and Expire CO_2

Pulmonary and alveolar ventilation aerate the lungs and maintains the partial pressures of CO_2 and O_2 in alveolar air. The diffusion of both respiratory gases is governed by several factors that are included in the Fick's law of diffusion. Both CO_2 and O_2 diffuses down their partial pressure gradients. The two sites of diffusion are the alveolar capillary membrane in the lungs and at the microcirculation between the capillary blood and the cells.

The carrying capacity of blood for CO_2 and O_2 is both determined by the partial pressures of CO_2 and O_2. For oxygen, the content is associated with the major carrier, hemoglobin which is increasingly saturated as the PO_2 is increased. For CO_2, its content comprises three forms, as carbamino-hemoglobin, dissolved CO_2 and as plasma bicarbonate, the latter product of CO_2 hydration within erythrocytes, making two-thirds of its total blood content.

The red cell is unique in its central role not only in oxygen transport but also in moving metabolic CO_2 from the cells to the lungs. The hemoglobin (Hb) within erythrocytes binds oxygen and CO_2. Red cell hemoglobin is also a buffer for hydrogen ions and this reaction is part of the pathway that promotes the generation of plasma bicarbonate, the major converted species of CO_2 in blood.

K.K. Mah and H.M. Cheng, *Learning and Teaching Tools for Basic and Clinical Respiratory Physiology*, SpringerBriefs in Physiology, DOI 10.1007/978-3-319-20526-7_2

Two red cell enthusiasts gave their names to effects that integrate the key role of hemoglobin in both CO_2 and oxygen transport. The Bohr effect describes how the hemoglobin-oxygen saturation and thus the O_2 blood content is influenced by the blood PCO_2/pH. The Haldane effect on the other hand (or the other Hb hand) states that the CO_2 content in blood is affected by the amount of oxygen dissolved in blood or the PO_2.

Carbon dioxide physiologically is more than merely a metabolic waste product. CO_2 is a major determinant of blood pH, which is closely guarded as neurons especially are sensitive to extracellular fluid (ECF) acidity/alkalinity. The respiratory system is thus a major organ that collaborates with renal functions to maintain a narrow range of ECF pH. Tissue carbon dioxide is by wonderful design also a metabolite vasodilator and this vascular action nicely enables tissues to provide for itself a greater blood perfusion during higher metabolic activity (active hyperemia). The cerebral and coronary vasculature also exploits the CO_2 vascular resistance-lowering actions in autoregulaton to maintain a relatively constant blood flow.

The pulmonary circulation represents the entire cardiac output from the right ventricle and the mixed venous blood from the whole body is delivered to the lung for re-oxygenation. Obviously, the pulmonary arterial blood does not serve to supply oxygen to the lungs and the unique vascular response of pulmonary vessels to hypoxia (vasoconstriction) highlights this contrast with hypoxic vasodilation in all other systemic arterioles. To accommodate the great volume of cardiac output, the pulmonary vessels are also relatively compliant and mechanical vasodilation is produced by increased pulmonary arterial pressure that distends and recruits pulmonary vessels during increased physical activity.

The systemic and pulmonary vascular networks are thus physio-logically designed to ensure adequate cellular respiration and optimal lung oxygenation together with the expiration of CO_2.

2.1.1 Ventilate, Oxygenate V/Q Matchmake

One of the best models to help students appreciate the different meanings of ventilation and oxygenation in lung function is to use the differential matching and consequences of ventilation and blood perfusion in the normal upright lung. The students would have learnt that alveolar Ventilation is better at the basal regions of the lung compared with the apical alveoli. More of each tidal volume flows to refresh the basal portions of the upright lungs.

Students who are observant would also notice and be puzzled that the partial pressures of oxygen however rises progressively from the base to the apex of the lungs. Clearly, alveolar ventilation alone cannot account for the alveolar air partial pressures of O$_2$ and CO$_2$ (the PCO$_2$ is also less in the apical alveoli although these are less ventilated than the basal alveoli).

Oxygenation refers to the uptake of oxygen from the alveoli into the pulmonary capillaries. Net oxygenation is achieved when the PO$_2$ in pulmonary blood equalizes the alveolar air PO$_2$. If we then consider not merely oxygenation from the aspect of PO$_2$ equilibrium across the alveolar-capillary membrane, but Rate of oxygenation (as vol O$_2$/min), then it might surprise the students that the oxygenation Rate is higher at the base compared to the apex of the lungs although the apical alveoli has higher PO$_2$.

At this juncture, the tutor can summarize by stating

1. that ventilation alone does not determine the value of alveolar air partial pressures of respiratory gases
2. Alveolar air PO$_2$ and PCO$_2$ are dependent on the ventilation and perfusion matching.
3. Lung Oxygenation to reach PO$_2$ equilibrium and Rate of lung oxygenation (ml O$_2$/min) should be distinguished.

The explanation for the better Rate of oxygenation at the base of the lungs is the greater pulmonary blood flow. This is the place where the students can be asked to discuss how this higher pulmonary blood flow benefits oxygenation Rate when understood from the term 'perfusion-limited oxygenation' that normally takes place in the lungs.

2.1.2 The F Notes

The upper and lower F notes. This respi octave between the two F notes symbolizes the diffusive Functions at the lungs and at the tissues. The letter F specifically points to the Fick's law that describes the factors governing diffusion.

The most essential factor is the partial pressure gradients for both oxygen and carbon dioxide. If the lungs represent the 12 and the cells the 6 on a clock face, the overall diffusive direction of oxygen is clockwise from lungs to cells and for CO_2, also clockwise from cells to the lungs.

Secondly, the total surface area available for diffusion is also important. In the lungs, this area represent the combined alveolar epithelial membrane and the apposing pulmonary capillary endothelial cell surfaces. Deeper tidal volume increases the alveolar area and both vascular recruitment and distention increase the exchange area of the capillary side of the alveolar-capillary interface.

At the tissues, total surface area is increased solely by changes in capillary hemodynamics. Arteriolar vasodilation e.g. during exercise increases the capillary hydrostatic pressure in muscle tissues and this distends and recruits non-fully patent parts of the microcirculation. Besides the increased in exchange area, the diffusion distance between the blood and the cells in proximity is decreased.

In the alveolar-capillary (a-v) 'membrane', the thickness of this membrane is not altered significantly in normal persons. However in certain lung pathophysiology, called 'alveolar-capillary block', the a-v membrane is thickened and diffusion impairment expecially for oxygen occurs. Non-equilibration of oxygen results and arterial PO_2 is reduced.

F for Fick is also associated with Adolf Fick who gave a Principle that is useful to estimate regionl blood flow. For pulmonary blood flow determination, which is also eh cardiac output, Fick's principle that applies the law of conservation of mass, makes use of blood oxygen as the parameter measured in mixed pulmonary venous, arterial blood and the normal rate of O_2 uptake or oxygenation in the lungs.

2.1.2.1 Respiratory Rhythms Are the Music of Life (Key of F)

The note F is for Fick, the last name of Adolf Fick who is known for his Fick's Principle and Fick's law of diffusion.

The Fick's Principle is a general principle that utilizes the concept of conservation of mass to formulate an equation to determine regional blood flow. For pulmonary blood flow, the parameter that is measured is oxygen content in the pulmonary arterial and pulmonary venous blood, and the rate of uptake of oxygen by the pulmonary blood. This incident is also a measure of cardiac output that is pumped by the right ventricle.

Fick's Law of diffusion describes the multiple factors that affect the rate of simple diffusion of uncharged substances. For oxygen and CO_2, the main factor is the partial pressure gradient for both the gas exchange at the alveolar capillary membrane and at the cellular level, the diffusion from microcirculation into the ISF and into the cells (for oxygen) and in the opposite direction for CO_2.

The other important major Fick's diffusive factor is surface area for diffusion. E.g. Recruitment of capillaries at both the lungs and the tissues expand the available area for diffusion of the respiratory gases.

Diffusion distance is another component in the Fick's law of diffusion. The thickness of the alveolar capillary is fairly constant in health but is significantly increased in 'alveolar capillary block' in certain pulmonary diseases and in pulmonary edema. At the tissues, recruitment of capillaries not only increase the surface are for diffusion but also shorten the diffusive distance for gas exchange.

There is also a permeability factor for O_2 and CO_2 in Fick's equation for diffusion. This is seldom a problem for both gases at the lungs and the tissues. Like the acoustics in a music hall that provide good transmission of sound, the membrane permeability for both O_2 and CO_2 is excellent.

2.1.2.2 Respiratory Rhythms Are the Music of Life (Key of C)

The keyboard note of C, in particular middle C is a reference point in the several octaves of music that can be played out. The C is for Content of blood gases, oxygen and CO_2. This is also a physiologic reference point for the purpose of the respi-cardiovascular system. The normal arterial blood Content of oxygen at 20 ml O_2/100 ml of blood ensures a rich reservoir of oxygen in circulation. At rest, the arterial blood Content of O_2 is depleted only about 25 % due to tissue extraction.

The arterial blood Content of CO_2 is also controlled (this might be a surprising fact if CO_2 is merely viewed as a metabolic product). CO_2 blood Content, is determined by the partial pressure of CO_2 which is a component of the major bicarbonate/PCO_2 buffer system in ECF. Life depends on a homeostatic maintenance of ECF pH, the internal environment of all living cells in the body.

In this context, the key note of C can remind the student that it is also CO_2 that acts as the primary chemical regulator of normal respiration.

2.1.3 The C Notes, Content of CO_2, O_2

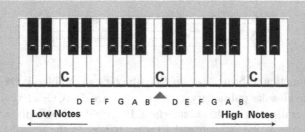

The **C notes. C is for Content** of oxygen and carbon dioxide in blood. The arterial blood has a higher O_2 carrying capacity than venous blood. Conversely, the venous blood carries more CO_2 than arterial blood.

It might be surprise students to be told that in arterial blood, the total volume for CO_2 in the blood is higher, more than twice the total oxygen content. The value is ~20 ml O_2/100 ml blood and for CO_2 it is 48 ml CO_2/100 ml blood. For O_2 Content, the total is very much dependent on a Carrier while for CO_2, the CO_2 carrying capacity is made possible by a CO_2 hydrating enzyme, Carbonic Anhydrase (C@).

Both the Contents of oxygen and CO$_2$ are finally dependent on their partial pressures.

For oxygen, PO$_2$ determine the hemoglobin saturation with O$_2$ and the total O$_2$ content is ~99 % in the form of Oxyhemoglobin.

For CO$_2$ blood Content, the PCO$_2$ determine the degree of production of bicarbonate from the enzyme, C@-catalysed hydration of CO$_2$. BiCarbonate constitute ~70 % of the total CO$_2$ content in blood.

The student should note that de-oygenated venous blood has still 75 % of the blood O$_2$ content in arterial blood i.e. 15 ml O$_2$/100 ml blood.

For CO$_2$, it might surprise the student to C and note that arterial CO$_2$ content is still 92 % of the venous blood CO$_2$ content! Venous blood CO$_2$ content is 52 ml CO$_2$/100 ml blood compared with CO$_2$ in arterial blood at 48 ml CO$_2$ %. This high Content of CO$_2$ in blood tells us that CO2 has other physiologic roles and this gas is not simply a metabolic by product to be removed by respiration. Indeed, CO$_2$ in solution in body fluids, forming carbonic acid is a component of the major chemical buffer system, that together with bicarbonate anions, maintain the pH of the extra-cellular fluid.

Play me a C …. CO$_2$, Content, C@

Play me another C …and we will also note that the Contents of both O$_2$ and CO$_2$ are maintained and regulated by sensors called C hemoreceptors, found in the brain stem and at specific vascular locations.

2.1.4 Interface, Surface in Cardio-Respiratory Physiology

The lungs function to bring 'face to face' the air and the blood. Oxygenation occurs at the alveolar-pulmonary capillary interface. De-CO$_2$ of venous blood also occurs at the same air-blood interface. Passive diffusion is the respiratory event that governs the oxygen and CO$_2$ exchanges at the alveolar-capillary interface. Fick's law of diffusion details the determinants of diffusion rate, that includes the major factor of area for exchange. The role of alveolar ventilation is to maintain a normal partial pressure of O$_2$ and CO$_2$ so that there will be an optimal gas gradients for the diffusion of both O$_2$ and CO$_2$.

The bloody river of life connects the air-blood lung interface above with the destination, the Interface between the blood and the interstitial/intracellular fluid. The diffusion of towards cells and metabolic CO$_2$ into the capillary blood again follow their respective available partial pressure gradients.

Along the journey from lungs to cells and cells to lungs, the Surface of the red cell hemoglobin serves a crucial function in both the transport of O$_2$ and CO$_2$. Oxygen binds as oxyhemoglobin and CO$_2$ interact with hemoglobin as carbamino-hemoglobin and also indirectly as protonated—Hb, the protons generated from the hydration of CO$_2$.

On the surface of hemoglobin, the oxygen and CO$_2$ play out their affinity game with the results of their game always benefitting the audience at both the Interfaces. The two referees at the Interfaces are Mr Bohr and Mr Haldane. The gas ('Guest') in the alveolar and cellular audience at both the interfaces are refreshed and re-junevated by the hemoglobin games between oxygen and CO$_2$.

2.1.5 Bloody Multitasker Red Cell

The simple red cell that has lost its nucleus is not that simple after all! The red cell has taken on multiple other functions essential for life and health.

Firstly, its main cytosolic resident, the hemo-Globin carries oxygen to meet the needs of all nucleated cells for their metabolism. The hemo-Globin binds co-operatively to O$_2$, up to four O$_2$ per HGb, each subsequent O$_2$ binding more avidly than the precious O$_2$. This special characterictic of HGb allows it to be 90 % saturated with O$_2$ even at the low partial pressure of O$_2$ at 60 mm Hg. The normal PO$_2$ in oxygenated arterial blood is about 100 mm Hg. Thus HGb not just carries O$_2$ but the way it specially and intimately transport O$_2$ protects the cells in our bodies from hypoxia.

The hemo-Globin also has the essential role in transporting carbon dioxide from the cells to the lungs. The hemo-Globin (hGb) co-operated with another cytosolic resident, an active catalytic individual called Carbonic Anhydrase (C@). The C@ speeds up the hydration of metabolic CO$_2$ that then generates bicarbonate and protons. HbG buffers the protons and this promotes the formation of bicarbonate.

The buffering of protons by HGb is also a part of the overall mechanism for blood pH control. Venous blood is just slightly acidic thanks to HGB proton removal functions. The HGb itself is also able to accommodate a small fraction of the CO$_2$ to form carbamino-HGB. The binding of HGb for CO$_2$ has another simultaneous unique action in enhancing the unloading of oxygen to the cells (Bohr's effect).

The HGb also binds nitric oxide (NO). YES, the HGB binds NO!. Nitric oxide is a vasodilator and this action on vascular smooth muscle contributes to the greater blood perfusion in metabolically active cells.

What a bloody multi-gifted red cell...

Oxygen transport...

Credit Prof Fizzo and student Fizzee, by Saw Shier Khee, medical graduate of Universiti Malaya 2014

2.1.5.1 The Telephone and Respiratory Numbers to Recall

In Human Physiology, there is a long list of quantitative values that can overwhelm the student. Some values have a wide range e.g. hormonal blood concentrations and I normally discourage students from memorizing them other than to note their low nanomolar or micromolar levels in plasma.

The partial pressures of both oxygen and CO_2 however are universal values and are similar in all persons irrespective of gender or age. This constancy in PO_2 and PCO_2 is the basis of analytical diagnosis made by determining arterial blood gas values. The blood partial pressures together with blood pH and bicarbonate concentration allows identification of respiratory or metabolic causes pH disturbances in the patients.

The values of Oxygen and CO_2 starting from the lungs to arterial blood to tissue level to venous blood and back again to the pulmonary alveolar capillary completes the circuit of gas transport. This clockwise changing values of PO_2 and PCO_2 can be easily recalled by assigning a telephone number to oxygen and CO_2. The direction of dialing the telephone starts at 12 O' clock. Thus the oxygen call number is 103-97-40-40 representing respectively, values at the lungs, arterial blood, microcirculation at the cells and venous blood. CO_2 number is then 40-40-46-46.

2.1.6 Dial CO_2

The telephone as we know it with the dial. The dial turns clockwise as we call up a number. The clockwise movement reminds me of the systemic circulation, from the left heart pumping out oxygenated blood to the venous return of de-oxygenated blood into the right atrium/ventricle.

The movement of CO_2 is determined by the partial pressure gradient of CO_2. The 4 points of the clockwise circuit (N, S, E, W) can be assigned a number/value representing the PCO_2 at that point. The values will represent the PCO_2 in alveolar air, arterial blood, tissue, and mixed venous blood respectively. The PCO_2 at these locations will be 40, 40, 46, 46 mm Hg respectively.

So to aid storing in our memory the PCO_2 as the bllod moves from the lungs to the cells and back to the lungs, we can dial the CO_2 number 4040 4646.

We can add on the operator number, 0 (zero) to give **0 40 40 46 46**. The zero reminds us of the fresh, inspired air in each tidal volume where the PCO_2 is ~0 mm Hg. The alveolar air PCO_2 at 40 mm Hg is a dynamic equilibrium value. This stable alveolar PCO_2 is determined by the ratio of the cellular CO_2 production to the alveolar ventilation.

The expired tidal volume is a mixture of about two thirds alveolar air and one-third unexchanged inspired air that remains in the anatomical dead space. Thus the expired air PCO_2 has a value between 40 (alveolar) and 0 (inspired air) mm Hg PCO_2.

You call up CO_2 as you would call up a friend. Carbon dioxide is not merely a metabolite waste, an enemy of the body. CO_2 participates in several essential events in homeostasis. CO_2 is a major component of the main buffer system in the extracellular fluid, the bicarbonate/carbonic acid buffer. Metabolic CO_2 is also a vasodilator and this accounts for some of the chemical mechanisms for vascular autoregulation of blood flow to brain and the heart and local tissue hypercapnia also contributes to active hyperemia.

CO_2 is also an enhancer of oxygen unloading from hemoglobin to the cells (Bohr's effect). To send a message for more oxygen, the cell can transmit **Oxygensupply@Bohr.com**

2.1.7 Exhale and Confess

1. I believe that the alveolar PCO_2 is kept relatively constant by a balance between alveolar ventilation and rate cellular CO_2 production.
2. I believe that CO_2 is the main chemical regulator of normal breathing
3. I believe the small partial pressure gradient of less than 10 mm Hg for optimal CO_2 exchange at the tissues and in the lungs matches the greater sensitivity of chemoreceptors for CO_2.
4. I believe that CO_2 is not merely a metabolic byproduct or waste from cells
5. I believe that CO_2 is an essential component of the main ECF buffer system to control pH.
6. I believe that CO_2 vasodilates arterioles during autoregulation of cerebral and coronary blood flow when blood pressure falls.
7. I believe that the red cells contain the essential key carbonic anhydrase enzyme, without which the transport of metabolic CO_2 in venous blood from the cells to the lungs would be inadequate.
8. I believe that metabolic CO_2 binds hemoglobin and enhances the release of oxygen from oxyhemoglobin.
9. I believe that exercise hyperventilation is primarily to maintain arterial PCO_2 rather than to increase blood oxygen content.
10. I believe that it is more than a wonderful coincidence that expired CO_2 from our respiratory tree becomes the main source substrate source of photosynthesis in trees and plants which then release oxygen for our human respiratory functions.

2.1.8 H^+Aldane Homeostatic Teaching

HbALDANEffect@Hotmail.com

I write this on the board at the beginning of a tutorial class. This physiologic email always brings a smile to students. Several aspects of carbon dioxide transport are highlighted to give handles for grasping the essentials of CO_2 physiology. These include

1. The main chemical regulator of respiration is CO_2 and not oxygen.
2. The transport characteristic of CO_2 transport does not saturate as for hemoglobin-O_2 interactions.
3. The operating partial pressure grandient of CO_2 diffusion at the cells and at the alveolar capillary membrane is much narrower, 40/46 mm Hg compared to the diffusion gradient for PO_2 (~100/40 mm Hg).
4. The Hb in the HbALDANE reminds the students that hemoglobin is the same key actor that participated in the Bohr Hb-O_2 diffusion drama.
5. HbALDANE's effect refers to the way the oxygenation status of hemoglobin alters the affinity of Hb for both protons and CO_2.
6. Oxygenated blood physio-logicallly has less capacity to carry CO_2 than de-oxygenated blood draining the tissues.
7. The students are asked to think outside the CO_2 metabolic waste 'box' and review if CO_2 has any physiologic actions.
8. This question intends to direct the student to recall that local tissue regulation of blood flow involves CO_2 as a vascular arteriolar dilator
9. The respiratory system has three inter-related functions, carry oxygen, transport CO_2 and participates in pH control. In pH regulation CO_2 takes on the role of a major component (carbonic acid) of the key bicarbonate/carbonic acid buffer system in the ECF.
10. And Haldane seeks to transport CO_2 from cells to the lungs. The affinity rights of Haldane are interlinked with the affinity rights of Bohr! Haldane claims that de-oxy Hb binds CO_2 more while Bohr declares that CO_2/H^+ binding to Hb promotes more de-oxy Hb which is saying that more O_2 is beneficially unloaded to the cells.

2.1.9 Pulmonary Vascular Resistance (Pvr)

I write on the board

$$BP = \mathbf{CO} \times TPR$$

$$pulm\,bp = \mathbf{CO} \times pvr$$

The cardiac output (CO) in bold in the two equations have the same value, representing the right and left ventricular cardiac outputs.

Since the students will know that the pulmonary circulation is a low blood pressure circuit, the pulmonary blood flow, equivalent to the cardiac output must be possible only because it is served by a low resistance pulmonary vascular resistance.

I develop this discussion further by asking the students to look at the two equations of systemic and pulmonary hemodynamics.

In the first systemic formula, the students knows that the CO and the TPR are determinants of systemic arterial blood pressure. I put forward Are both CO and pvr then also determinants of pulmonary arterial pressure?

This question requires the students to consider the relatively complained pulmonary blood vessels which are easily distensible by increased intra-vascular pressure. This suggests to the students that in the pulmonary circulation, an increased blood pressure can reduce the pvr (occurs by vascular distention and recruitment).

The students are reminded that this 'BP causing vascular resistance changes' is not a significant phenomenon in the systemic circulation. The systemic arterioles are also innervated by vasoconstrictor sympathetic fibers which provide some of the vasomotor tone.

This contrasts with the insignificant role of autonomic nerves in regulation of pvr.

The peculiarities of pvr is further expanded to illustrate the contribution of vascular transmural pressure in altering pvr.

1. Changes of pvr during a respiratory cycle can be explained by the inverse changes in transmural pressure in the alveolar capillaries and in the extra-alveolar capillaries during inspiration and expiration.
2. Changes in vascular transmural pressure that decreases in a vertical upward direction accounts for the lower perfusion of apical alveoli, as the alveolar air pressure compresses the capillaries. The driving pressure at the apex of the upright lungs becomes the difference between arterial and alveolar pressure, rather than the venous pressure.

Finally, there is also the unusual hypoxic-induced vasoconstriction of pulmonary vascular smooth muscles, in contrast to the vasodilation in other regional tissues when local hypoxia occurs. Pulmonary hypoxic increased resistance has to do with balancing ventilation and perfusion and this vasoconstriction does not obviously serve an appropriate metabolic function.

Chapter 3
Regulation of Respiration

Credit Physio female character, 'Phyrettie', by Anabella Diong, Final year medical student, University of Malaya

Abstract Life, breathing and oxygen naturally go together. Certainly, breathing is partly controlled by the demand for oxygen in cellular metabolism. Unconscious breathing is an automatic, involuntary process set by pacemaker neurons in the respiratory centers in the brain stem. Breathing is unique in that it can also be under volition by breath-bolding or taking a deep breath. In the latter voluntary effort, cerebral neurons send action potentials that bypass the brain stem respiratory 'pacemakers' to the inspiratory muscles.

© The Author(s) 2015
K.K. Mah and H.M. Cheng, *Learning and Teaching Tools for Basic and Clinical Respiratory Physiology*, SpringerBriefs in Physiology, DOI 10.1007/978-3-319-20526-7_3

27

Keywords Hypercapnia · Hypoxia · Chemoreceptors · Respiratory cycle · Alveolar ventilation

3.1 Introduction: Ready to Regulate and Alveolar Ventilate

Sensory receptors for oxygen and carbon dioxide, chemoreceptors transmit afferent signals to the respiratory neurons in the brain stem. There are two distinct chemoreceptor groups, the central brain stem chemoreceptors in the vicinity of the pacemaker neurons and the arterial peripheral chemoreceptors in the carotid/aortic bodies. Physiologically, it is the CO_2 rather than the oxygen that provides the primary regulating stimulus for respiration. The homeostasis of CO_2 in the body is associated with the tight control of extracellular pH and neuronal functions are particularly sensitive to pH in the cell milieu.

Thus, conceptually, increasing respiratory rate and tidal volume during hyperventilation of exercise is not completely geared towards oxygenation and delivery to cells. A crucial purpose of increasing alveolar ventilation during physical activity is to maintain a relatively constant PCO_2 in the ECF/blood. Thus alveolar ventilation changes are tied to cellular CO_2 production in order to achieve a stable alveolar air and consequently arterial PCO_2.

Exercise hyperventilation only marginally increases the blood oxygen content; this is due to the near complete saturation of hemoglobin at rest and the more determining role of increased cardiac output that provides the greater rate of O_2 supply to the tissues.

The key role of CO_2 is also evident from the particular sensitivity of the cerebral arteriole to CO_2. Local increase in cerebral CO_2 will vasodilate the blood vessels in order to increase the perfusion and remove the excess metabolic CO_2. A voluntary hyperventilation illustrates this effect as the induced hypocapnia produces cerebral vasoconstriction with resulting dizziness.

Hypoxi is not necessarily always associated with hypercapnia. If we breathhold, decreased O_2 and increased CO_2 do occur concurrently. Any causes of hypoventilation also leads to hypoxia and hypercapnia.

If the cells are oxygen deprived due to sluggish blood, the stagnant hypoxia means more O_2 can be extracted from each unit of blood flow and the venous PO_2 can be expected to decrease. However, in stagnant hypoxia, it is not definite for the blood CO_2 to be elevated. Similarly, in anemic hypoxia, since normally the tissue O_2 extraction from capillary blood is only 25 %, there is a potential reservoir of O_2 in blood to ensure adequate uptake of O_2 to meet cellular needs. Hypercapnia is not inevitable in anemic hypoxia.

The students might recall that in the special situation of hypoxic hypoxia of high altitude, the hypoxia is associated with a respiratory alkalosis and hypocapnia. In heavy exercise, the lactic metabolic acidosis stimulates respiration to produce a respiratory alkalosis. However, in moderate exercise, it remains a physiologic mystery to be unraveled as to how hyperventilation is sustained since the measured blood PCO_2 and PO_2 are relatively unchanged at 40 and 97 mm Hg respectively.

3.1.1 Hold Your Breath

holdbreathmoment

The moment you hold your breath

A number of physiological events occurs.

Firstly, you have just shown that breathing can be under voluntary control. Here, neurons in the cerebral cortex override the spontaneous respiratory pacemaker activity in the brain stem. Breathing is temporarily halted by the voluntary action.

Secondly, the alveolar partial pressures of respiratory gases begin to change. The PO_2 decreases as uptake of oxygen continues into the pulmonary blood in the absence of fresh ventilated air. For PCO_2, it increases as diffusion of CO_2 from mixed venous blood proceeds in the absence of ventilation. The alveolar PCO_2 is determined by the ratio of the cellular CO_2 production to the alveolar ventilation.

The length of time endurable during breath-holding is not actually due to the increasing hypoxia as the partial pressure of O_2 decreases in the blood. CO_2 is the main chemical regulator of respiration and the potent stimulus of the respiratory centers via the arterial chemoreceptors and directly on the central chemoreceptors. Thus the 'break-point' is reached with the rise of PCO_2 to the threshold, when the desire to breathe again becomes overwhelming and breath-holding is released!

The primacy of CO_2 can be expanded by noting that the drop in PO_2 from a normal 100 to 60 mm Hg only reduces the hemoglobin-oxygen saturation by 10 %. The need to expel rising CO_2 from the blood has to do with the sensitivity of the neurons to pH. Carbonic acid forms from dissolved CO_2 and lowers the ECF pH.

The primacy of CO_2 control is also illustrated by the greater need to remove CO_2 during physical activity by increased ventilation. Hyperventilation is not the main factor in supplying more O_2 to the cells during exercise. Rather it is the increased cardiac output. The hyperventilation is aimed at maintaining the arterial PCO_2 by stabilizing the alveolar air PCO_2.

So when we voluntarily hyperventilate before going underwater, our enjoyable time spent in aqueous environment is due to the induced hypocapnia. A longer time underwater, breath-holding is made possible before the hypercapnic drive takes over.

3.1.2 Hypoxia and PCO₂ Too?

Quite often in students' answer scripts, I read the sentence 'when hypoxia occurs, the partial pressure of CO_2 also rises...'
This sentence assumed loosely and imprecisely by students is a good platform to review in more depth the concept of subtypes of hypoxia. By going through each causative category of hypoxia, the students can judge for themselves whether PCO_2 invariably will be elevated during hypoxia.

(a) Hypoxic hypoxia: by definition, the partial pressure of arterial PO_2 is decreased. Any causes of hypoventilation will lower the arterial PO_2 and raise the PCO_2. Ventilation/perfusion mismatch also produce a decreased PO_2 and a higher PCO_2. So yes, in hypoxic hypoxia, we can generally say that the PCO_2 is elevated.

(b) Histotoxic hypoxia: If the cells are unable to use blood oxygen fully, then less metabolic CO_2 will be generated. Thus in cases of metabolic inhibition, the histotoxic hypoxia is not associated with a rise in blood PCO_2.

(c) Anemic hypoxia: The oxygen carrying capacity is reduced due to decreased hemoglobin function, either a reduced concentration or presence of abnormal hemoglobin. In this situation, the PO_2 is unchanged. Although the total O_2 blood content is reduced, the cells can extract more O_2 from arterial blood since the usual O_2 tissue extraction is just 25 % of the blood O_2 content. So the production of CO_2 should not be affected to cause any rise in blood PCO_2.

(d) What about in Stagnant hypoxia? The blood does not circulate well in stagnant hypoxia due to e.g. weak cardiac pump function. Again, the arterial blood PO_2 is unchanged since lung oxygenation is normal and it is only the O_2 delivery rate (ml O_2/min) to cells that is inadequate. The students should be able to reason that the cells can extract a greater portion of the blood O_2 content from the sluggish capillary flow, to meet their metabolic needs. Therefore, there is then no inevitable increased in CO_2 addition to the blood.

Perhaps students have a general impression from the condition of increased physical activity when lack of (greater demand for) oxygen and increased CO2 are associated. This analytical class exercise into the causation of hypoxia should emphasize to the students to be exact and precise in use of physiologic terms and to qualify and specify mechanisms so that physiologically inaccurate general statements are avoided.

3.1.3 Respiration Recap

1. The 'negative' in 'negative' pressure normal breathing refers to the sub-atmospheric (ref atm pressure at 0 mm Hg) alveolar pressure that is generated in order for the inspired tidal volume air to flow into the lungs.
2. Elasticity does not mean stretchability or distensibility in physiology. An elastic structure like the lungs has elastic recoil that Opposes stretch or distention.
3. Ventilation is volume of air/min and Oxygenation is volume of O_2/min.
4. The apex of the upright lungs has less alveolar ventilation from each tidal volume but relative to the blood perfusion, the apical alveoli are 'over-ventilated' (V/q is higher).
5. The alveolar air PO_2 and PCO_2 are not due not on alveolar ventilation alone but more correctly are dependent on the ventilation/perfusion (V/Q) matching.
6. The partial pressure of oxygen in plasma is only due to the dissolved O_2 and the amount of oxygen bound to hemoglobin does not affect the PO_2 in solution.
7. Normal chemical control of breathing is due predominantly to CO_2 and not oxygen. Only in chronic hypercapnia in lung patients does the chemical control switch to a 'hypoxic drive' due to de-sensitization/adaptation of the sensory chemo-receptors.
8. All the local changes in the exercising muscle physio-Logically enhances unloading of oxygen from hemoglobin (reduces Hb–O_2 affinity). These changes include higher temperature, lower pH and increased PCO_2, all of which shift the Hb–O_2 curve to the right (increased P_{50} index).
9. Red blood cells are essential for transport of CO_2 in blood because it contains the enzyme carbonic anhydrase (C@) and the hemoglobin which binds CO_2 and which also buffers hydrogen ions produced by the C@ action.
10. The pulmonary vascular resistance is decreased directly by increased pulmonary arterial pressure due to pressure distention and recruitment. This unique cause and effect mechanism is due to the relatively compliant pulmonary blood vessels, and this pressure/resistance effect is not obvious in the systemic circulation.
11. A 'shunt' in respiratory physiology refers to a geograPHYic situation where the blood is not oxygenated. A physiologic shunt is due to normal anatomical arrangement of some of the bronchial and cardiac circulations. An intra-pulmonary shunt is where the local regional alveolar ventilation is reduced (v/Q decreased e.g. blocked airway). What is called a right to left shunt refers to cardiac septal defect that allows de-oxygenated blood short-circuited into the left chamber (s).
12. Exercise hypeventilation does not serve primarily to increase the blood oxygen content as the cardiac output is the dominant provider of increased O_2/min to the cells. This is explained by the perfusion-limited oxygenation at the alveolar-capillary membrane and the near saturation of hemoglobin even at normal arterial PO_2 (sigmoid Hb-O_2 graph).
13. Exercise hyperventilation is however essential to remove the increased metabolic CO_2. The alveolar air and thus the arterial PCO_2 is determined by the ratio of the cellular CO_2 production to the alveolar ventilation.

3.1.4 Respiration Rewind

3.1.4.1 Respimetaphys

1. Increased in ventilation during exercise is not primarily to increase oxygen supply but for increased removal of CO_2 that is produced at a higher rate from the cells
2. Tissue CO_2 is both a vasodilator and an Enhancer of oxygen unloading from hemoglobin
3. CO_2 as carbonic acid, regulated by respiration, is also a component of the main chemical buffer system (bicarbonate/carbonic acid) in the ECF.
4. The relatively low pulmonary vascular resistance sustains the resting pulmonary blood flow which is the whole cardiac output from the right ventricle.
5. The relatively compliant pulmonary vessel s accommodate the increased cardiac output during exercise by vascular distention and blood vessels recruitment.
6. The alveolar partial pressure of oxygen and CO_2 is dependent not just on ventilation but on the matched ventilation—perfusion relationships.
7. Increased blood flow (cardiac output) and not a higher blood oxygen content is the primary provider of increased oxygen delivery rate to the cells during increased tissue metabolism.
8. Carbon dioxide and not oxygen is the primary chemical regulator of respiration.
9. Oxygen dissolved in plasma determines the partial pressure of oxygen in blood and thus the percentage saturation of hemoglobin-O_2 but not the Hb–O_2 affinity.
10. All local tissues changes during increased metabolism (temperature, acidity, CO_2) favour oxygen extraction by reducing the Hb–O_2 affinity.
11. Alveolar Ventilation is simply the volume of fresh air movement into lungs (liters/min) but Oxygenation is the volume of O_2 moved by diffusion (ml O_2/min) across the alveolar-capillary membrane.
12. Partial pressure of O_2 and CO_2 in blood is in equilibrium with Oxyhemoglobin and Carbamino-hemoglobin respectively.
13. The partial pressure of O_2 in arterial blood is determined not by the concentration of hemoglobin but by the alveolar air PO_2 and the minimal physiologic shunt.
14. The recoil of the lungs, expanded since birth, is always inwards whereas the chest wall recoil is inwards or outwards depending on the lung volume.
15. Pulmonary blood flow is not purposed to supply oxygen for the metabolic needs of lung tissues!
16. Pulmonary blood flow and equalized and synchronized with systemic cardiac output via simultaneous systolic contractions of right and left ventricles and operation of the intrinsic Starling's cardiac muscle mechanism in the two rhythmic ventricular pumps arranged in series.

17. The right atrial pressure is functionally equivalent to the central venous pressure and fluctuates during a normal respiratory cycle. The mechanics of normal breathing actively lowers the alveolar air pressure to become sub-atmospheric (negative with ref to atm = 0 mm Hg) and tidal volume of air enters the lungs down the pressure gradient.

18. The mechanics of normal breathing actively lowers the alveolar air pressure to become sub-atmospheric (negative with ref to atm = 0 mm Hg) and tidal volume of air enters the lungs down the pressure gradient.

19. It makes physiologic sense for airway resistance to be lower during lung expansion and this is due to the bigger airways transmural pressure as the intra-pleural pressure is more negative with greater lung volumes.

20. It is good physiologic design that equilibration of PO_2 during oxygenation occurs very rapidly in less than half the capillary transit time as this makes possible more delivery of O_2 (ml O_2/min) to cells when the pulmonary blood flow increases.

21. The extraction of O_2 by the tissues at rest is only a quarter of the total blood O_2 delivered and this means there is abundant O_2 reserve for when a greater O_2 demand occurs.

22. The minimum drop of $Hb–O_2$ to 90 % when the Partial pressure is reduced considerably to 60 mm Hg is a protective feature of hemoglobin against hypoxia besides allowing for high altitude adventures to be enjoyed!

23. The shifting, inverse relationship between hemoglobin for O_2 and CO_2 at the cells and the lungs (Bohr's, Haldane's effect) demonstrates a co-operative interaction between the two respiratory gases on the erythrocyte carrier protein.

24. Elastic structures have elastic recoil and this opposes stretch and distention. This means that lung elasticity is physiologically inversely related to the lung compliance.

25. Surface tension force is an inward recoil force, tending to collapse the lungs. Pulmonary surfactant equalizes the effect of surface tension between small and larger alveoli and established alveoli stability.

26. You can hold your breath but not your heart beat.

27. All non-respiratory ('metabolic') causes of acid-base disturbance are compensated by changes in alveolar ventilation mediated and effected by chemoreceptor responses to pH and PCO_2.

28. Pulmonary vascular smooth muscle is uniquely constricted by hypoxia and this is purposed to minimize local ventilation-perfusion imbalance where the alveoli are relatively under-ventilated.

29. A trans-mural pressure is the pressure gradient across the 'wall' and is a distending or expanding pressure calculated from the difference between the inside and outside the 'wall' (airway, alveolus, blood vessels)

30. Inspiration is active, produced by action potentials from respiratory center neurons to respiratory muscles while expiration is passive, due to recoil of lung/chest wall.

31. The alveolar PCO_2 is dependent on the balance between the rate of cellular CO_2 production and alveolar ventilation. Note that alveolar PCO_2 is also said to be determined by the V/Q matching at the alveoli.

3.1.4.2 Swing and Cycles of Respiratory values

There are several values that fluctuate in a repetitive periodic cycle. From a low value to a higher value and then back to the starting minimum value. This swing in respiratory values are part of the mechanisms involved in the mechanics of breathing and in the gas transport to tissues. A few examples illustrate this

(a) the intra-pleural pressure from -3 to -5 to -3 mm Hg. At rest and during normal breathing the intra-pleural pressure (Pip) remains negative. The act of active inspiration via contraction of respiratory muscles makes this Pip more negative. This is essential to increase the trans-pulmonary pressure that then expands the alveoli. During expiration, the passive recoil of the chestwall/ lungs returns the Pip to its pre-inspiratory negative value.

(b) associated with the swing in the Pip values is the accompanying fluctuations in alveolar air pressure from 0 to -1 to 0 mm Hg. The zero mm Hg is with reference to atmospheric pressure. Normal breathing is termed negative pressure breathing, the 'negative' referring to the alveolar air pressure decreasing below 0 mm Hg. Inspired air moves into the lungs due to the pressure difference between the atmosphere and the alveoli. During normal expiration, the elastic recoil of the chest wall/lungs unit compresses the alveoli and the intra-alveolar pressure increases back to its pre-inspiratory value of 0 mm Hg.

(c) the partial pressure of oxygen swing from high 100 mm Hg in arterial blood to a less than half at 40 mm Hg in mixed venous blood. This cycle of PO_2 values are due to the two sites of conversion, oxygen extraction at the tissues (100–40 mm Hg) and re-arterialization or re-oxygenation of venous blood at the alveoli respiratory units.

3.1.5 Spring Respi-playground Swing

The Swing in the playground is a good place to review several 'swing' aspects of respiratory physiology.
Swing represents a cycle of repeated activity. Obviously, left to its own momentum after the initial energy push, the swing will gradually slows

down to a standstill. Cycles and rhythms in human physiology are sustained by energy in the 'pacemaker' or cycling cells. There are two essential cycles of cell activity critical for life, the heart beat and its cardiac cycle and the respiratory cycle.

We can view the respiratory cycles from different perspective.

1. There is the inspired tidal volume followed by the expired tidal volume.
2. The inspired tidal volume in turn is achieved by cycle of pressure changes in the intra-pleural space (intra-pleural pressure).
3. The intra-pleural pressure cycles to a more negative value during inspiration, from its sub-atmospheric mm Hg at the end of a normal expiration in the previous breath.
4. The breathing rate is determined by a cycle of action potential generation by brain stem respiratory neurons. 10 pulses of action potentials to the respiratory muscles will then produce a respiratory rate of ten breaths (10 inspirations) per minute.
5. The alveolar pressure also cycles from 0 mm Hg from the beginning of an inspired tidal volume to a sub-atmospheric value at mid-inspiration, returning to 0 mm Hg at the end of inspiration.
6. The cardiac cycle of the right side of the heart is reflected in the vascular pressure changes during systole/diastole in the pulmonary artery that perfuses the lungs.
7. Besides these cycles of physical, mechanical parameters, the circadian rhythms that affects hormone secretions also has some impact on lung function. In particular, for asthmatics, the relatively greater hyper-responsiveness of the lungs at night could be due to the fluctuations of the anti-inflammatory hormone, cortisol, which peaks near waking moment.

3.1.5.1 See-Saw, and Balance in Respiratory Physiology

Normal values are maintained in respiratory mechanisms by the inputs of factors exerting their opposing effects like weights on each side of the playground see saw. Thus the analogy of the playground instrument is a useful visual or mental tool to communicate the Physiology as educators. A few key sites in the respiratory system where the normal values are the result of balancing factors include

(a) the alveolar air partial pressure of oxygen, P_AO_2. This normal value of 103 mm Hg is the equilibrium value between alveolar ventilation that brings into the lungs, fresh, inspired O_2 and the continual uptake of oxygenation process that occurs across the alveolar capillary membrane into the pulmonary circulation,

(b) likewise for the alveolar partial pressure of CO_2, P_ACO_2. This is described in the Alveolar Ventilation equation where the alveolar PCO_2 is determined by the ratio of the cellular CO_2 production (V_{CO2}) over the rate of alveolar ventilation, V_A. Thus in the See-Saw picture, if the CO_2 production is the same but alveolar ventilation is increased (hyperventilation), the balance is now weighted towards a reduced alveolar air PCO_2. This is the picture of respiratory alkalosis.

3.1.6 Respi-Playground See-Saw

Spring See Saw balance

The see saw is the place where we all as children, have enjoyed playing our balancing games. In respiratory physiology, there are several moments where we observe essential balance between opposing factors. We can review and think of several here

1. the opposing recoil forces of the lung and chest wall at the end of a normal expiration,
2. the normal alveolar value of PO_2 and PCO_2 is a balance between fresh air input from inspired air and unceasing O_2 removal from or CO_2 addition to the respiratory alveoli.
3. Net diffusion of respiratory gases reaches a balance point when the partial pressures are equalized on each side of the exchange area at both the alveolar—capillary membrane and at the tissue blood-ISF, cellular interface.
4. If there is greater cellular demand for oxygen during physical activity, the situation of demand-supply is balanced several ways (i) increasing cardiac output (ii) vasodilation at tissue level (iii) enhanced unloading of O_2.
5. Ventilation is matched with perfusion in the normal lung. Any imbalance in ventilation/perfusion will produce a reduction in arterial PO_2.
6. The chemoreceptors function to maintain the alveolar air and arterial PCO_2 and PO_2. Any shift in the parameters is sensed and the partial pressure balance restored by either compensatory hyper- or hypoventilation.
7. The regulation of pH involves several buffer and also organ systems. The key buffer, bicarbonate/carbonic acid system, is intimately linked to the balance between renal and respiratory functions.

3.1.7 Dizzy with Respi

Our breathing can be voluntarily increased, both the depth and the frequency. We can recall doing this at the start of a sporting event.

Hyperventilation might be expected to increase oxygen supply to our cells and we make the voluntary breathing effort to ensure this in advance of the physical activity. This however might not achieve its O_2 purpose.

Prolonged hyperventilation in fact can begin to cause us to feel dizzy. This giddiness is not due to reduced blood oxygen content. The dissolved oxygen in blood is increased but the hemoglobin–O_2 saturation, that constitute ~99 % of total O_2 content is not increased significantly since at normal arterial PO_2, the Hb–O_2 is already ~98 % saturated. This dizziness is due to a decrease in cerebral blood flow and this then does reduce the delivery rate of O_2 supply to the brain. The giddiness can be viewed as a signal to alert us to stop the voluntary hyperventilation.

The reason for the cerebral effect is vasoconstriction of the cerebral arterioles that is produced by respiratory alkalosis, brought on by the hyperventilation.

Hyperventilation causes alkalosis due to a reduction in alveolar air PCO_2. The latter is determined by the balance/ratio between cellular CO_2 production and alveolar ventilation. This is represented in the Alveolar Ventilation equation, $PCO_2 \sim VCO_2/V_A$. The arterial PCO_2 and thus the blood CO_2 content is dependent on the alveolar air value.

The resulting respiratory alkalosis can be appreciated from looking at the change in the major ECF buffer system, the bicarbonate/carbonic acid. Increased in carbonic acid component will reduce the ratio that determines pH to less than the normal ratio of 20:1.

Hyperventilation prior to exercise does have the effect of increasing cardiac ventricular filling due to the 'respiratory pump' effect. This is effected as the intra-thoracic pressure that is reduced with deeper breathing decreases the central venous/right atrial pressure. A greater venous filling would then increase the cardiac output even before the exercise.

The physical effort of hyperventilating will increase general sympathetic activity. Adrenal catecholamine (adrenaline/noradrenaline) secretion will be released by sympathetic stimulation. Circulating adrenaline will act on the airways to produce bronchodilation.

Respiratory alkalosis will tend to lower free plasma calcium as albumin binds more calcium. In patients with 'irritable' neurons, the hypocalcemia-associated hyper-excitability of neurons can trigger seizures. Hypocalcemic tetany is the other hyperexcitability effect on skeletal muscles.

Chapter 4
Respiratory Recreation

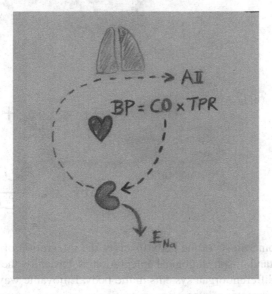

Doodling on recycled paper: Lingkage of lungs with kidneys and heart in blood pressure control

© The Author(s) 2015
K.K. Mah and H.M. Cheng, *Learning and Teaching Tools for Basic
and Clinical Respiratory Physiology*, SpringerBriefs in Physiology,
DOI 10.1007/978-3-319-20526-7_4

Credit Adlina Athilah Abdullah

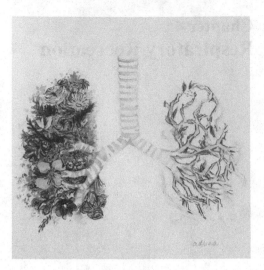

Respi-renal regulation of ECF pH

$$pH = pK_a + \frac{HCO_3^-}{PaCO_2}$$

$$pH \propto \frac{Kidneys}{Lungs}$$

Abstract The homeostasis of the body involves the integration of different organ systems. Understanding physiological integration is basically the ability to join the dots linking different organ systems in the body. Enjoyable ways to revise and think through homeostatic integration is to devise creative games such as card games and use of lyrics in songs.

Keywords Acid-base regulation · Partial pressures · Respi-CARDology · Renal compensation · Respiratory songs

4.1 Introduction: Ready to Integrate and Enjoy Respiratory Physiology

The homeostasis of the body, the 'wisdom of the body' is a collective wisdom and involves the integration of different organ systems in physiologic control. The lungs and the kidneys are linter-linked functionally in the regulation of ECF/blood pH.

The lungs remove the major carbonic acid product of metabolism while the kidneys excrete the non-carbonic acid daily load. The normal urine is generally acidic.

In addition, the lungs are involved indirectly in the long term control of arterial blood pressure. The kidneys are the key players in the physiologic daily drama of blood volume/blood pressure control. This renal role includes the secretion of renin that then triggers the angiotensin-aldosterone cascade of reactions to produce sodium retaining hormonal actions in the kidneys.

Total body sodium is the major determinant of ECF/blood volume. Here, the conversion of angiotensin I to its bioactive form angiotensin II occurs primarily as blood course through the pulmonary vasculature whose endothelia cells possess the angiotensin converting enzyme.

Renal conservation of sodium/volume balance also requires the neural action of the renal sympathetic nerve (rsn). The rsn reduces filtered sodium load, contributes the renal component of total peripheral (arteriolar) resistance and also directly increases proximal tubular reabsorption of sodium. Activation of rsn as part of the general effector sympathetic discharge is via reflex loop from baro and volume receptors during hypovolemia/hypotension. The volume sensors are present in the pulmonary vasculature besides the cardiac atrial wall.

Another physiologic integration is cardio-respiratory functions. The 'cell customer' is always right. The lungs provide the oxygenation of venous blood but blood must circulate before the O_2-replenished blood can reach its intended cellular targets. The scenario of stagnant hypoxia when cardiac pump failure occurs emphasized the essential, combined cardio-respiratory blood flow-airflow relationships.

Understanding physiological integration is basically the ability to join the dots linking different organ systems in the body. One enjoyable way to revise and think through homeostatic integration is to devise creative games. Some examples in this chapter are given. Card games can be used and I have called these Respi-CARDology. In respiratory physiology, there are considerable values that are quite essential to take note. Remembering these numbers e.g. pertaining to partial pressures, blood content can be aided by playing Respi-CARDology. Explaining cardio-respiratory functions with reference to specific normal values makes the physiology more precise and engaging.

The Clubs in each set of cards do resemble the pulmonary alveoli and innovative questions can be formulated using these cards. The symbol for perfusion in physiology is the letter 'Q'. So students can use the Queen of Clubs in card games to represent pulmonary blood flow. Another respiratory recreation that can be tried for those musically Inspired is to write the physiology to popular tunes or rhymes and a few attempts by the author are included in this chapter also.

4.1.1 A Lung with the Kidneys; the Kid Never Walk Alone

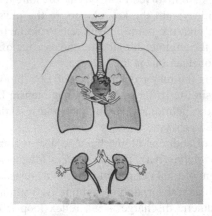

Both the lungs and the kidneys perform excretory functions. Metabolic carbon dioxide is expired and the kidneys secrete and excrete a variety of organic metabolic byproducts.

Beyond these dual excretory functions, the lungs and the kidneys are physiologic partners for life.

One life sustaining function by the lungs/kidneys is the maintenance of ECF/ blood pH. The key component that unites the lung and kidney in this integrative role is carbon dioxide. By this we are saying that CO_2 is not merely a metabolic byproduct. The lungs/kidneys are both linked in energizing the most quantitatively important buffer in the ECF. This is the bicarbonate/carbonic acid buffer system and CO_2 is thus a key chemical reaction constituent in this buffer system.

The lungs oxygenate the pulmonary blood flow and maintains a normal arterial blood oxygen content. If the lungs under ventilate or there are problems in diffusion at the alveolar–capillary membrane, the arterial PO_2 and O_2 content will be reduced. The kidneys will faithfully compensate for this hypoxia by secreting more erythropoietin. This hormone from the kidneys will circulate to the bone marrow and stimulate erythropoiesis. A secondary polycythemia is produced and this enables more uptake of oxygen in the lungs.

Not so obvious is the kidneys/lungs relationship in blood pressure regulation. This organ collaborative function is centered on the circulating peptide, angiotensin II. AII, a circulating peptide is generated from angiotensin I by the pulmonary endothelial enzyme, angiotensin converting enzyme (ACE). In turn, AI is derived from the initial secretion of renin from the kidneys. The renin-angiotensin system is the major hormonal regulator of arterial blood pressure. AII is both a potent vasoconstrictor that increases TPR and AII is also a major stimulator of adrenal aldosterone. Aldosterone maintains sodium balance via its action in the kidneys.

Sodium balance is tied to conservation of blood volume and hence long-term blood pressure control.

We can say that the kidneys prevent excessive over expansion of ECF. This is achieved by the kidneys' role in both sodium and water balance, the latter synonymous with osmoregulation. If overexpansion of ECF leads to pulmonary edema, exchange of respiratory gases will be impaired in the lungs. Here is a situation where kidney problems results in lung dysfunction.

4.1.2 Respiratory CARDology

During small group discussions and tutorials, there is much opportunity to employ innovative tools to engage the students to think through Physiology. Use of playing cards as stimuli for interaction among students is one way. The Clubs resemble the functional pulmonary unit, the alveoli. The numbers of the Club cards drawn or distributed can be a trigger or cue to ask questions relating to respiratory Physiology. To illustrate, the two Clubs can serve as an engaging point to review a variety of events or mechanism in respiration that have two major determining factors. A partial list of two component s that effect or affect respiratory processes include

- the Lung Recoil by Elasticity and Surface tension
- the total lung capacity comprising Inspiratory capacity and Functional residual capacity
- total pulmonary vascular resistance, a sum of alveolar and extra-alveolar vessels in series
- Carbon dioxide transport by hemoglobin in the form of carbamino and protonated hemoglobin
- Rate of oxygen delivery to the cells determined by product of cardiac output and blood O_2 content.
- partial pressure of a gas calculated from the fractional concentration and the total pressure.
- mechanical vasodilation of the compliant pulmonary vessels that produce recruitment and distention of pulmonary vasculature
- Ascertaining the normally of the physiologic shunt by comparing alveolar air and arterial blood PO_2.
- Respiratory control of blood pH via its link to the two base and acid components in the bicarbonate/carbonic acid buffer system in the ECF.

A creative extension of the use of playing cards can make us of the Hearts. Together the Clubs and the Hearts can be combined in play to formulate questions to talk and think on various physiologic cardio-respiratory homeostatic events.

4.1.3 Respi-CARDology, Six of Clubs

six of clubs. The respiratory system together with cardiovascular function adjusts for the increasing need for oxygen during physical activity.

The hemoglobin-oxygen saturation only begins to decrease more steeply below PO_2 of 60 mm Hg. This is seen in the sigmoid shape of the HB-O_2 graph and the curve plateaus out after **60 mm Hg**. Also, at 60 mm Hg, the Hb-O_2 is already 90 % saturated.

The main chemo control of breathing is blood CO_2. The hypoxic stimulus for ventilation becomes more significant only **below 60 mm Hg**. This hypoxic threshold for stimulation of arterial chemoreceptors matches the more precipitous reduction of Hb-O_2 saturation when the PO_2 drops below 60 mm Hg.

Normal 'negative' pressure breathing during Inspiration can be explained, for clarity, in **6 sequential steps**: (1) contraction of inspiratory muscles/diaphragm with action potentials from brainstem, (2) decrease in intra-pleural pressure, (3) increase in trans-pulmonary/trans-alveolar wall pressure, (4) expansion of alveoli, (5) decrease in alveolar air pressure to sub-atmospheric ('negative'). (6) Tidal volume of air moves into lungs.

Carbon dioxide transport from cells by RBC can also be viewed in **6 physio-chemical steps:** (1) CO_2 diffuses down PCO_2 gradient. (2) CO_2 dissolved in plasma and enters RBC. (3) Some CO_2 forms carbamino-Hb. (4) Most of CO_2 in RBC is hydrated by carbonic anhydrase and chemically forms protons and bicarbonate from carbonic acid. (5) The proton is buffered by deoxy-Hb (Haldane's effect) and this enhances the kinetics of HCO_3 generation. (6) Bicarbonate exchanges with chloride at RBC membrane and enters plasma as the major form of CO_2 transported to the lungs. (7) At the lungs reversal of the above CO_2 blood chemistry occurs.

4.1.4 Seven of Respiratory Alveolar Clubs

Seven of clubs.

A few parameters with seven can be noted in respiratory physiology.

- CO_2 partial pressure is 46 mm Hg in venous blood (some texts have it as 47 mm Hg)
- The CO_2 arterial–venous difference is much smaller (only **6–7**) mm Hg compared to that for oxygen (100/40 mm Hg)
- the water vapour pressure at body temperature of 37 °C is **47 mm Hg**
- The arterial blood PO_2 is **97 mm Hg** (marginal decrease from alveolar PO_2 of 105 mm Hg) due to physiologic shunt.
- The hemoglobin-O_2 saturation in arterial blood of PO_2 97 mm Hg is **97 %**
- The affinity index for Hb-oxygen binding is expressed with reference to the PO_2 required to produce 50 % Hb-O_2 saturation. The value is **27 mm Hg (P_{50} index)**.
- The Hb-O_2 saturation drops only 25 % after tissue extraction of O_2 and venous blood has Hb-O_2 saturation of **75 %** at PO_2 40 mm Hg.
- respiratory gas exchange begins at the **17th generation** of the bronchial tree (of life). The respiratory bronchioles branches from 17th to 19th generations…the final alveolar sacs is the 23rd generation.
- The **FEV_1/forced VC** ratio assesses airway resistance and in normals should be **70 %** and above.
- There are >7 lung space parameters that are part of common respi language; Tidal, Inspiratory reserve, Expiratory reserve, Residual volume and 4 lung capacities; Functional residual, Inspiratory, Vital and Total lung capacity.

4.1.5 Behold RespiPhysiology

Behold PhylosoPHY

1. Behold how the intra-pleural pressure is always negative at rest
2. Behold how the intra-pleural pressure becomes more negative during inspiration
3. Behold how breathing can be held voluntarily
4. Behold CO_2, the primary chemical regulator of breathing
5. Behold how the alveolar partial pressures of O_2/CO_2 is dependent not just on ventilation but on the matching between alveolar ventilation and pulmonary perfusion.
6. Behold the manner in which the PO_2 equilibrate very rapidly at the alveolar capillary membrane, which accounts for why oxygenation is described as 'perfusion-limited'.
7. BEHOLD how appropriate that the local changes in CO_2, pH and O_2 during increased cellular metabolism all enhance the unloading of O2 from hemoglobin to the cells.
8. BEHOLD how appropriate that the local changes in CO_2, pH and O_2 during increased cellular activity all vasodilate arterioles to increase blood flow to the more metabolically active tissues.
9. Behold that the dissolved CO_2 is in equilibrium with carbamino-Hb, with protonated Hb and also with the hydration reaction that generates plasma bicarbonate.
10. Behold that the x-axis of the hemoglobin-O_2 dissociation curve is not arterial PO_2 but just PO_2.
11. Behold that the dissolved oxygen that gives the PO_2 is in equilibrium with oxy-hemoglobin
12. Behold that even in venous blood, de-oxygenated Hb still has 75 % oxygen.
13. Behold that in arterial oxygenated blood, carbamino-Hb is still present.
14. Behold how chronic hypercapnia de-sensitizes the chemoreceptors and the patient's breathing becomes sustained and shifted instead on a 'hypoxic drive'
15. Behold that the recoil force of the lungs, tending to collapse the alveoli is due to both the elastic recoil and the surface tension force.
16. Behold that the trans-mural pressure across the alveolar wall is also termed the trans-pulmonary pressure.
17. Behold that there are fluctuations of heart rate (higher in Insp), central venous pressure (lower during Insp) during a respiratory cycle. (also pulmonary vascular resistance)
18. Behold that the major change I pulmonary vascular resistance is not mediated by nerve or hormones but by sheer mechanical, blood pressure distention which also recruits pulmonary blood vessels.
19. Behold the recoil of the lungs is always inwards whereas the recoil of the chest wall depends on the position of the chest wall.
20. Behold that the beta receptor binding that produce bronchodilation are mainly due to circulating adrenaline and not sympathetic neuro-transmitter.

21. BEHOLD the pulmonary vascular response to hypoxia is vasoconstriction and not vasodilation and this serves the purpose of ventilation/perfusion balance rather than metabolism.

10 of clubs…there are quite many tens or multiples of 10 in pulmonary physiology.

We breathe around 10–12 breaths per minute.

Dry, atmospheric inspired air has ~16 × 10 (160) mm Hg PO_2 and humidified inspired air has ~15 × 10 (150) mm Hg

The anatomical dead space in a 70 kg male adult is ~15 × 10 (150) ml

Arterial partial pressure of oxygen is approximately 10 × 10 (100 mm Hg).

The arterial oxygen content is approx 2 × 10 ml (20) O_2/100 ml of blood

Mixed venous blood PO_2 is 4 × 10 (40) mm Hg.

The unique hemoglobin interaction with O_2 protects the cells from hypoxia; when the PO_2 drops from 100 to 60 mm Hg, the Hb-O_2 saturation is only decreased by ~10–90 % saturation.

The resting O_2 usage by cells is 25 × 10 (250) ml/min

Arterial blood partial pressure of carbon dioxide is 4 × 10 (40) mm Hg

However in terms of content CO_2, being more soluble in plasma, arterial CO_2 content is actually more than that for oxygen, ~5 × 10 (50) ml CO_2/100 ml blood (specifically 48 in arterial and 52 in venous)

and the resting rate of CO_2 production by cells is 20 × 10 (200) ml CO_2/min

4.1.6 Respiratory Songs; No Lungs, No Song

Physiology notes to Respiratory Lyrics

1. Alvin Alveolar singing to Carmen Capillary

Pulmonary surfactant has the essential function in lowering surface tension force in all the alveoli and this acts to maintain alveoli stability. In surfactant deficiency, the surface tension force causes an increased recoil of the alveoli with result-ing collapse of the smaller alveoli. The Work of breathing is increased with the reduced lung compliance.

This inward recoil can exert a force that pulls fluid into the interstitial space from the alveolar capillary. In serious conditions, fluid can also enter the alveolar space. Pulmonary edema interferes with diffusion and exchange of O_2 and CO_2 as the diffusion distance becomes lengthened by the accumulating fluid.

2. Joy to my Lungs

This Christmas Carol has lyrics that highlight the essential pressure changes that allow normal breathing to occur. The four pressures to describe in normal 'nega-tive' pressure breathing starts with a drop in intra-pleural pressure. The trans-pul-monary pressure is the transmural pressure across the alveolar wall. Also called trans-pulmonary pressure, this distends the alveoli and the intra-alveolar pressure becomes sub-atmospheric. Air then enters the lungs along the atmospheric-alveo-lar pressure gradient during inspiration.

3. Oxy to Lungy

This respi song explains the unique response of the pulmonary vessels to oxy-gen changes. Unlike systemic arterioles, blood vessels in the lungs demonstrate a hypoxic pulmonary vasoconstriction (hpv). This hpv does nor serve a metabolic function but is directed to optimizing ventilation/perfusion matching in the nor-mal lungs. However in the generalized hypoxia of high altitude, continua exposure to the low oxygen pressure can result in a generalized hpv. Pulmonary hyperten-sion is produced and this will present the right ventricle with an increased cardiac work. Compensatory right heart hypertrophy takes place.

In mountaineers, compensatory hyperventilation improves the blood satura-tion and content of oxygen. A side effect is respiratory alkalosis which is resolved partially be renal excretion of more bicarbonate during acclimatization over a few days. The kidneys increases its erythropoietin secretion in response to the hypoxia. A secondary polycythemia helps to sustain an adequate oxygen delivery to the cells.

4. My Bonnie lies over the Ocean

This popular tune is now lyricized with respiratory language pertaining to the unique features of oxygen transport. Hemoglobin-bound oxygen represents the bulk (99 %) of totoa blood oxygen content. Thus the affinity of Hemoglobin (Hb) for oxygen is central to meeting the metabolic demands of cells. More active cells release more CO_2 and the local pH of the cells are also more acidic. By providen-tial design, both increased CO_2 and lower pH decreased the Hb-oxygen affinity (a P_{50} affinity index is used where the 50 indicates 50 % achievable at a PO_2 in mm Hg). From the cellular perspective, a reduced Hb-O_2 affinity benefits the cells as more oxygen is unloaded. This enhancing effect of CO_2/hydrogen ions was described in 1904 and credited to the Danish physiologist, Christian Bohr.

1. Alvin Alveolar singing to Carmen Capillary

Don't Cry for Me Argentina
Don't Cry for me, E-de-ma
Surface Tension, I Recoil from You
Fluid, into ISF stay
My poor Compliance
Lost my Surfactant
Long Diffusion Distance

2. Joy to my Lungs

Joy to my Lungs!
I can breath.
Inspire, Alveoli Expand.
Let Trans-mural Pressure Distends
Trans-Pulmonary Pressure Expands
Intra-Pleural Pressure Descends
Negative Breathing Depends
Inspire! Tidal Volume
I can Sing!

3. Oxy to Lungy

When I am down, PO_2 so weary
Pulm Pressure up, my Right Heart burden be
Hy-po-xic Pulm Va-so Constriction
Acclimatize, Hyperventilate for me
You Raise me up, so I can Stand on Mountains!
You raise me up, Alkalosis Res-pi
I am strong, EPO, polycythemia,
You Raise me up, to Saturate H-b.

4. My Bonnie lies over the Ocean!

My Bonnie Lies Over the Ocean
Scottish Folk Song
Traditional
Arranged by James Baird

My Bohr's is good for Oxy-gen
My Bohr's Unload O_2 to cells
CO_2, Oxygen Reduction
Low pH also Bohr's can you see?
Feed back O Feed back
Bohr's reduce Hb-O_2 aff-inity
Feed back O Feedback
High P_{50} is good for me!

Part II
Learning Scenarios

Chapter 5
Mechanics of Respiration

Abstract In this chapter, the reader is introduced to the application of the fundamentals of the mechanics of respiration discussed earlier in the first section of this book. Numerous scenarios requiring calculations are prepared in this chapter. It is hoped that if the reader are able to work them out conscientiously, the reader would have a deeper understanding of the mechanics of respiration.

Keywords Lung function test · Poiseuille law · Lung compliance · Laplace's law · Surfactant · Dead space

5.1 Introduction

The first scenario involves a lung function test with the objective of highlighting to the readers the clinical relevance of knowing the various lung volumes and capacities. There are not merely numbers to be memorised to pass examinations.

The clinical importance of compliance, surfactants and the Laplace's law is highlighted in two clinical scenarios. The consequences of altered lung compliances and impaired surfactant productions leading to various diseases are discussed.

Dead spaces and its consequences to alveolar ventilation are extensively covered as these are topics that often confuse students. The various dead spaces: anatomical, physiological and apparatus dead space are explained with problems requiring calculations in the scenarios.

© The Author(s) 2015   53
K.K. Mah and H.M. Cheng, *Learning and Teaching Tools for Basic
and Clinical Respiratory Physiology*, SpringerBriefs in Physiology,
DOI 10.1007/978-3-319-20526-7_5

A scenario on the clinical application of the Poiseuille Law is introduced in the hope that readers will appreciate how and why the diameter and length of tubes can determine the resistance of gas flow and subsequent the work of breathing.

5.1.1 Theme: Lung Function Test

Mr A, a 60 years old man was admitted because of wheezing and coughing. He has asthma for the past 40 years. On examination he has a barrel shaped chest and had severe wheezing over both lungs. Chest X-ray revealed bilateral hyper-inflated lung fields. He was nebulised with salbutamol and his wheezing subsided.

Results of his lung function test are as follows (estimated normal values in brackets)

Vital capacity: 3000 ml (4600 ml)
Inspiratory capacity: 2000 ml (3500 ml)
Functional residual capacity: 4000 ml (2300 ml)
Total lung capacity: 5900 ml (6000 ml)
Tidal volume: 700 ml (500 ml)
Expiratory reserve volume: 1000 ml (1100 ml)
Residual volume: 3000 ml (1200 ml)

Question:

(a) Calculate his inspiratory reserve volume
(b) Explain the results of his lung function test
(c) What is the effect of nebulised salbutamol on this patient?

Answer:

(a) His inspiratory reserve volume is 1300 ml (Inspiratory capacity minus tidal volume = 2000 − 700 ml)
(b) His residue volume is very high. This is due to air trapping in the lower respiratory tract by dynamic airway compression during excessive forceful expiration in the asthmatic patient. This is reflected in his hyper-expanded barrel chest as well as radiological findings of hyper-inflated lung fields.

Functional reserve volume is the sum of residue volume and expiratory reserve volume. The large functional residue capacity is this patient is due mainly to the enlarged residue volume in patients with air trapping. Expiratory reserve is fairly unchanged here.

His calculated inspiratory reserve volume (1300 ml) is very low compared to the estimated normal of 3000 ml. This is not unexpected as in asthmatic patients as well as any patient with chronic obstructive airway diseases, the lungs are already almost maximally over inflated at the beginning of inspiration.

The low vital capacity in is a result of the markedly reduced low inspiratory reserve volume seen this type of patients.

(c) Salbutamol has selective β_2-receptor agonist properties which causes dilatation of bronchial smooth muscle. This relieves the bronchospasm in asthmatic attacks. According to Poiseuille law which states

$$Q = \Delta P \pi r^4 / 8\eta l$$

And we know $Q = \Delta P/R$

Therefore $R = 8\eta L / \pi r^4$

And hence $R \propto 1/r^4$

Where Q = flow, ΔP = pressure drop, L = length, R is the resistance to flow, η = viscosity and r = radius of the tube.

Bronchodilation, which increases the radius of the airways, will lead to a significant reduction in the resistance to airflow in the respiratory tract $(R \propto 1/r^4)$.

When there is bronchodilatation, the patient will then be able to force more air out during expiration. As a consequence, there will be less air trapping, residue volume will decrease and expiratory reserve volume will increase.

5.1.2 Theme: Compliance

A lung function test on a 20 year old man with a functional residue capacity (FRC) of 2.5 L is shown below.

Question:

(a) What is lung compliance and how is this obtained from the graph below? Calculate the lung compliance.
(b) Calculate the specific lung compliance.
(c) Explain the clinical significance of compliance.

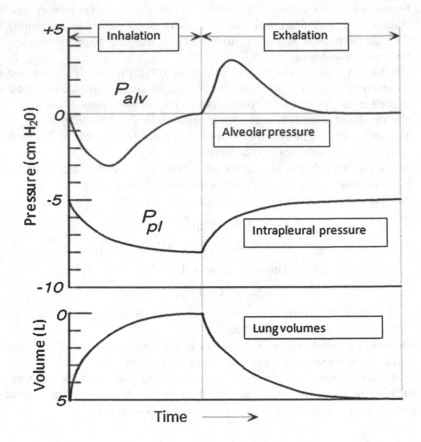

Answer:

(a) Lung compliance refers to the magnitude of change in lung volume as a
     result of the change in pulmonary pressure. Pulmonary compliance is calcu-
     lated using the following equation, where ΔV is the change in volume, and
     ΔP is the change in pleural pressure

$$\text{Compliance} = \Delta V / \Delta P$$

Graphically it refers to the slope of the pressure-volume curve at a particular
lung volume i.e. volume change per unit of pressure change ($mL/cmH2O$).

From the graph above, the person has a tidal volume of 500 ml and has an
intrapleural pressure before inspiration of -5 cm H_2O and -8 cm H_2O at the
end of inspiration.

Therefore his lung compliance is $= 500 \, ml/3 \, cm \, H_2O)$

$$= 167 \, ml/cm \, H_2O \, (normal \, value \, 200 \, ml/cm \, H_2O)$$

(b) His specific compliance = Compliance divided by FRC

$$= 167 \, ml/2500 \, ml/cm \, H_2O$$

$$= 0.067/cm \, H_2O \, (normal \, value \, 0.05/cm \, H_2O)$$

(c) Two major factors determine lung compliance. The first factor is the elasticity of the lung tissue. Any thickening of lung tissues due to diseases will decrease lung compliance. The second factor is surface tensions at the air water interfaces in the alveoli. This will be discussed in the next scenario.

Low compliance indicates a stiff lung and translates to extra work to effect a change in the intra-pleural pressure to inhale in a normal volume of air. This occurs as the lungs become fibrotic, congested or atelectatic and lose their distensibility Persons with low lung compliance tend to take shallow breaths and breathe more frequently.

In a highly compliant lung, as in emphysema, the elastic tissue is damaged. They have no problem inflating the lungs during inhalation but have extreme difficulty during exhalation. In this condition extra work is required to exhale air out of the lungs.

5.1.3 Theme: Surface Tension, Laplace Law

In an experiment, a young man was found to have an alveolar wall tension force (T) of 0.06 N/m. His left and right lungs are represented by 2 alveoli of different sizes with radius of 0.00004 m and 0.00008 m, respectively.
Question:

(a) What is surface tension?
(b) What is Laplace's Law and what is the collapsing pressure of the 2 individual alveoli?
(c) What is the consequence of this different collapsing pressure and how is this overcome?

Answer:

(a) The attractive forces between water molecules on the water–air interface are responsible for the phenomenon of surface tension. Below the surface of the liquid, each molecule is pulled equally in every direction by neighbouring water molecules, resulting in a net force of zero. The molecules at the surface however do not have other molecules on all sides of them above and below the surface and are therefore pulled into a spherical shape. This spherical shape minimizes the "surface tension" of the water–air interface according to Laplace's law.

(b) The air pressure (P) needed to opposed the collapsing force of surface tension (T) in an alveolus of radius r is expressed by the law of Laplace

$$P = 2T/r$$

The collapsing forces of both alveoli are

$$P = (2 \times 0.06 \text{N/m})/0.00004 \text{ m}$$

$$= 3000 \text{ N/m}^2$$

$$P = (2 \times 0.06 \text{N/m})/0.00008 \text{ m}$$

$$= 1500 \text{ N/m}^2$$

(c) From the calculations above, the alveoli with the smaller radius has twice the collapsing pressure of the larger alveoli. Since the two alveoli are connected, the smaller alveoli will collapse into the larger alveoli which become bigger.

However in normal circumstances, this does not occur. Surfactant, produced by Type 2 pneumocytes of the alveoli has the ability to decreases the alveolar surface tension. During expiration, surfactant molecules are packed tightly together, so that the surfactant molecules separate the water molecules and reduce the total alveolar wall tension. During inspiration when the alveoli radius is larger, the surfactant molecules are forced apart. The attractive forces of the water molecules located between the surfactant molecules will increase thereby increasing surface tension. In a way therefore, surfactants acts to regulate the size of the alveoli preventing them from getting to big or too small.

Pulmonary surfactant therefore reduces surface tension in the alveoli. Lung compliance is increased which allows the lung to inflate easily and decreases the work of breathing.

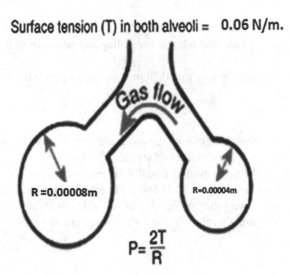

Surface tension (T) in both alveoli = 0.06 N/m.

R = 0.00008m   R=0.00004m

$$P = \frac{2T}{R}$$

5.1.4 Theme: Dead Space

A medical student's respiratory function and blood gas results are as follows

Parameters	Results
V_E	9 L/min
V_A	6 L/min
f (breaths)	12/min
P_aO_2	98 mm Hg
P_aCO_2	38 mm Hg
pH	7.4

Question:

(a) Explain why the minute ventilation, V_E, is different from the alveolar ventilation, V_A?
(b) What is 'Dead Space'
(c) How is 'Dead Space' calculated
(d) Describe a clinical example of dead space.

Answer:

(a) The difference is that the alveolar ventilation does not take into account the dead space ventilation, V_D. Alveolar ventilation only takes into account gases that are inhaled in and participate in gaseous exchange.
(b) The "Dead Space" of the lungs refers to those areas of the lung which do not participate in gas exchange. There are two types of dead space, anatomical dead space and physiological dead space.
Anatomical dead space is the volume of air that is inhaled into the conducting airways. These do not possess alveoli and any air inspired into those areas cannot be used for gas exchange.
The dead space volume derived from the anatomical dead space as well as any non-perfused alveoli is referred to as the "Physiologic Dead Space". Non-perfused alveoli whether pathological or physiological are also part of the dead space as any air entering such alveoli cannot efficiently be used for gas exchange.
(c) Dead space is calculated from the 2 equations below
V_E, the minute ventilation

$$V_E = V_T \times f, \quad \text{where } V_T \text{ is the tidal volume, f is the respiratory rate.} \quad (5.1)$$

V_A, the alveolar ventilation

$$V_A = (V_T - V_D) \times f, \quad \text{where } V_D \text{ is the dead space volume.} \quad (5.2)$$

From Eq. 5.1:

$$V_T = V_E/f$$
$$= 9/12$$
$$= 0.75\ L$$

From Eq. 5.2:

$$V_A = (V_T - V_D) \times f$$
$$6 = (0.75 - V_D) \times 12$$
$$V_D = 0.75 - 6/12$$
$$= 0.25\ L$$

(d) In anaesthesia, when the patient is intubated and ventilated, the endotracheal tube and the anaesthetic circuit constitute what is called the "Apparatus Dead Space".

5.1.5 Theme: Ventilation

The medical student in the scenario above then goes snorkeling. Assuming each centimetre length of the snorkel has a volume of 10 ml.

Question:
How far down can he swim before there is no gas exchange in his alveoli?
Answer:
$V_T = 750$ ml
$V_D = 250$ ml.

Therefore the portion of the tidal volume available for gaseous exchange in alveolar ventilation, V_A, is $(V_T - V_D)$ or $(750 - 250) = 500$ ml.
When the snorkel dead space is equal to $(V_T - V_D)$ or 500 ml, there will be ZERO gas exchange.
Since each centimetre length of the snorkel has a volume of 10 ml.
The maximum depth that he can go underwater when no gaseous exchange occurs despite breathing in a tidal volume of 750 ml is (500/10) or 50 cm.

Question:
Predict his $P_a\,CO_2$ when he is at a depth of 25 cm.

Answer:
This can be determined from the Alveolar Ventilation Equation below

$$V'_A = \left(V'_{CO_2}/P_{aCO_2}\right) * K$$

V'_A = Alveolar Ventilation Rate
V'_{CO_2} = Cellular production rate carbon dioxide
P_{aCO_2} = Partial pressure of arterial carbon dioxide

K = Unit correction factor

From this equation, assuming the rate of carbon dioxide exhalation is constant, then

$$V'_A \propto 1/P_{aCO_2}$$

Or graphically

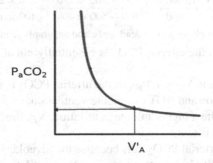

Therefore at a depth of 25 cm, his total dead space now equal to physiological dead space plus half the snorkel dead space, this is equal to $250 + (500/2)$ ml or 500 ml.

The inspired tidal volume that is available for gaseous exchange is now $(750 - 500)$ ml or 250 ml.

His effective alveolar ventilation drops to half or (250×12)ml or 3L/min.

From the graph and the equation $V'_A \propto 1/P_{aCO_2}$., his predicted P_aCO2 is approximately doubled or (38×2) or 76 mm Hg.

Question:
If he hyperventilates, will his P_aCO_2 decreases?

Answer:
Usually, an elevation of his P_aCO_2, will stimulate his central chemoreceptor causing him to hyperventilate in order to exhale excess CO_2 to maintain CO_2 homeostasis. However, whether the P_aCO_2 will change or not will depends on his respiratory pattern.

Question:
Consider this various respiratory patterns below when he is 25 cm below the surface.

Which respiratory pattern will increase or decrease his P_aCO_2?

Respiratory pattern	VT (ml)	f (per min)
a	600	20
b	750	20
c	900	10

Answer:

Pattern	V_T (ml)	f (per min)	V_E (ml/min)	$V_T - V_{D+App}$ (ml)	V_A (ml/min)	Predicted P_{aCO2}
Control	750	12	9000	$750 - 500 = 250$	$250 \times 12 = 3000$	76 mm Hg
a	600	20	12,000	$600 - 500 = 100$	$100 \times 20 = 2000$	Higher
b	750	20	15,000	$750 - 500 = 250$	$250 \times 20 = 5000$	Lower
c	900	10	9000	$900 - 500 = 400$	$400 \times 10 = 4000$	Lower

$V_T - V_{D+App}$ = Volume of physiological dead space and apparatus dead space

In a normal person, the arterial PCO_2 is essentially equal to and is determined by the alveolar PCO_2.

And from the equation $V'_A \propto 1/P_{aCO_2}$, the arterial PCO_2 is inversely proportional to the alveolar ventilation and NOT the minute ventilation.

In pattern 'a', despite a higher minute ventilation, V_E, the arterial PCO_2 rises as its alveolar ventilation is less.

In pattern 'b', the arterial PCO_2 falls because the alveolar ventilation is higher.

In pattern 'c', despite having the same minute ventilation, V_E, of 9L/min, the arterial PCO_2 falls because the alveolar ventilation is higher.

5.1.6 Theme: Poiseuille Law

Question:
A 30 year old male motor vehicle victim arrived with a crush injury to the chest. The doctor decided to intubate the patient as the patient was having great difficulty to breath. The nurse who is new handed the doctor a size 6 endotracheal tube but the doctor requested for a bigger size 8 endotracheal tube instead. After a successful intubation, the doctor cut the endotracheal tube so that only a shorter length protrudes from the lips.

(a) Explain why the doctor preferred a bigger size 8 endotracheal tube.
(b) Explain the reasons for the action of the doctor.

Answer:

(a) A size 6 endotracheal tube has an internal diameter of 6 mm whereas a size 8 endotracheal tube has an internal diameter of 8 mm. A larger diameter endotracheal tube will enable the patient to be ventilated more efficiently. This can be explained by the Poiseuille law which states

$$Q = \Delta P \pi r^4 / 8\eta l$$

And we know $Q = \Delta P/R$

Therefore $R = 8\eta l/\pi r^4$

And hence $R \propto 1/r^4$

Where Q = flow, R is the resistance to flow, η = viscosity and r = radius of the tube.

In this case, the resistance to laminar air flow in an endotracheal tube with a larger diameter will be less.

$$R \propto 1/4^4 \text{ is much less when compared with } R \propto 1/3^4$$

(b)   Poiseuille law in action again.

$$Q = \Delta P \pi r^4 / 8\eta l$$
$$\text{And } R = 8\eta l / \pi r^4$$

If η and r are constant, then $R \propto l$.

Hence a shorter endotracheal tube will have less resistance to laminar air flow.

In addition a shorter endotracheal tube will have less dead space (apparatus dead space) and will not contribute to the existing physiological dead space of the patient.

Chapter 6
Oxygen Cascade and Blood O_2 Content and Transport

Abstract The partial pressures at the various levels of the oxygen cascade are described here. Emphasis is placed in the application of the Alveolar Gas Equation in this chapter. Hopefully readers will have a clearer picture on how the amount of oxygen carried by the arterial blood differs under various conditions.

Keywords Oxygen content · Oxygen dissociation curve · Carbon monoxide · Oxygen cascade · Alveolar gas equation · (A-a) oxygen gradient

6.1 Introduction

The dissolved oxygen component in blood, often overlooked, will be appreciated by readers once they have gone through the relevant clinical scenarios here.

The oxygen dissociation curve (ODC) a favourite examination topic is discussed at length. Readers are often confused when the parameters of the 'y' axis in the ODC are changed from oxygen saturation to oxygen content.

The Alveolar-arterial oxygen gradient (A-a) O_2, is another topic which seems to instill fear in students. Three clinical scenarios are dedicated to this topic. The Alveolar-arterial oxygen gradient (A-a) O_2 is a very simple and useful diagnostic tool to help clinicians differentiate the various causes of hypoxemia especially when patients are on oxygen therapy.

6.1.1 Theme: Oxygen Content

Mr S, has a blood test done and his haemoglobin level is 12 g/100 ml blood.

© The Author(s) 2015
K.K. Mah and H.M. Cheng, *Learning and Teaching Tools for Basic and Clinical Respiratory Physiology*, SpringerBriefs in Physiology, DOI 10.1007/978-3-319-20526-7_6

Question:

(a) How much oxygen can a gram of haemoglobin carry? (a mole of haemoglobin has a molecular weight of 64.5×10^3 g).
(b) How much oxygen is the haemoglobin in Mr S carrying?

Answer:

(a) 1 mol of Hb has a molecular weight of 64.5×10^3 g
    1 mol of oxygen occupies a volume of 22.4×10^3 ml at standard temperature and pressure.
    But 1 mol of Hb can carry 4 mol of oxygen because each Hb has 4 heme groups, each carrying a mole of oxygen
    Therefore 1 mol of Hb can carry $(4 \times 22.4 \times 10^3)$ ml of oxygen
    Or 64.5×10^3 g of Hb can carry $(4 \times 22.4 \times 10^3)$ ml of oxygen
    Therefore 1 g of Hb can maximally carry $(4 \times 22.4 \times 10^3/64.5 \times 10^3)$ ml of oxygen or 1.39 ml of oxygen.
    This is also known as the oxygen carrying capacity of 1 g of haemoglobin.
    Most calculation for the oxygen carrying capacity of haemoglobin uses a value of 1.34 ml oxygen/g of Hb instead of 1.39 ml oxygen/g Hb.
    This is because the value of 1.39 ml oxygen/g Hb represents chemically pure haemoglobin.
(b) The total amount of oxygen carried solely by the haemoglobin of Mr S is $12 \times 1.34 = 16$ ml of oxygen/100 ml blood.

6.1.2 Theme: Oxygen Dissociation Curve

Two brothers, Mr A and Mr B were involved in a motor vehicle accident.
    Mr A suffered a concussion but Mr B was worse off with a fractured pelvis and significant blood loss.
    Blood results revealed that Mr A had a haemoglobin level of 15 g per dl and his brother haemoglobin of 7.5 g per dl.
    Assuming both had no lung or cardiovascular injuries.

Question:
What is the arterial haemoglobin oxygen carrying capacity of both brothers?

Answer:
The arterial haemoglobin oxygen carrying capacity of Mr A is the maximum amount of oxygen that can be carried by the haemoglobin in 100 ml of his blood if assuming his haemoglobin is 100 % saturated.
Since 1 g of Hb when fully saturated carries 1.34 ml of oxygen
A haemoglobin of 15 g per dl of blood will carry 15×1.34 or 20.1 ml of oxygen at 100 % saturation.
Similarly his brother Mr B will have an arterial haemoglobin oxygen carrying capacity of 7.5×1.34 or 10.05 ml of oxygen per dl of blood when his haemoglobin is fully saturated.

Question:

What would be the partial pressure of oxygen in the arterial blood of Mr A and Mr B if both are breathing air only?

Answer:

Assuming both have no lung injuries and both are breathing air with no supplemental oxygen, there should be no diffusion block in the transfer of oxygen from their alveoli to their pulmonary capillaries. The partial pressure of oxygen in the arterial blood of both would be equal to their alveolar oxygen partial pressure.

From the Alveolar Gas Equation, when breathing air at atmospheric pressure, the partial pressure of oxygen in the alveolar is about 100 mm Hg. Therefore both brothers will have an oxygen partial pressure in their arterial blood equivalent to almost 100 mm Hg.

The partial pressure of oxygen in the arterial blood is independent of haemoglobin levels.

Question:

What is the arterial oxygen content of both Mr A and Mr B if the pulse oximeter reading on both brothers shows a saturation of 98 %?

Answer:

The arterial oxygen content

= the amount of oxygen carried by the haemoglobin + oxygen that is dissolved in the plasma.

= 1.34 (haemoglobin × SaO_2) + (PaO_2 × 0.003)   where SaO_2 = Saturation of oxygen

The dissolved oxygen in Mr A's blood is derived from Henry's law,

Dissolved Oxygen/dl = Partial pressure of oxygen × solubility of Oxygen

$$= PaO_2 \times 0.003 \, ml \, O_2/dl/mm \, Hg$$

$$= 100 \, mm \, Hg \times 0.003 \, ml \, O_2/dl/mm \, Hg$$

$$= 0.3 \, ml \, O_2/dl$$

The dissolved component is the same for both the brothers.

For Mr A, the amount of oxygen carried in the arterial blood.

= the amount of oxygen carried by the haemoglobin + oxygen that is dissolved in the plasma.

= (20.1 ml of oxygen × 98 % saturation) + 0.3 ml O_2/dl

= 19.7 + 0.3 ml O_2/dl

= 20 mlO_2/dl

For Mr B, the amount of oxygen carried in the arterial blood.

= the amount of oxygen carried by the haemoglobin + oxygen that is dissolved in the plasma.

= (10.05 ml of oxygen × 98 % saturation) + 0.3 ml O_2/dl

= 9.8 + 0.3 ml O_2/dl

= 10.1 ml O_2/dl

Question:

Draw an Oxyhaemoglobin Dissociation Curve (ODC) for both Mr A and Mr B

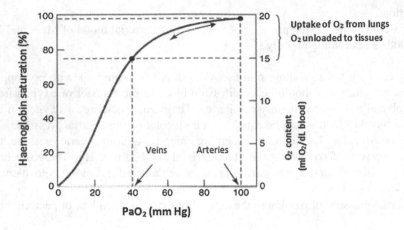

Answer:

On the Oxyhaemoglobin Dissociation Curve, when the 'Y' axis reads percentage saturation of haemoglobin, the curve for both brothers superimposes and therefore are the same.

Question:

Draw the Oxyhaemoglobin Dissociation Curve (ODC) for both Mr A and Mr B if the 'Y' axis now replaced by arterial Oxygen Content.

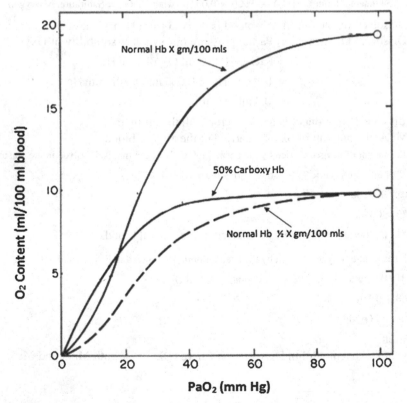

Answer:
Both have sigmoid curves but on the ODC, the 2 curves are different as both have different oxygen content level at the corresponding same PaO_2 level.

Note: This is a common mistake made by students in examinations when they are instructed to draw the oxygen haemoglobin dissociation curves when the parameters for the 'y' axis are changes from the 'usual' oxygen saturation to oxygen content. Hopefully this learning scenario will be able augment students understanding of the ODC.

6.1.3 Theme: Oxygen Carrying Capacity 1

A third unrelated patient Mr C is now admitted to the same hospital. He is diagnosed to be suffering from carbon monoxide poisoning.

Question:
Assuming that his haemoglobin is 15 g/dl and his haemoglobin is 50 % saturated with the carbon monoxide, what would be the maximum oxygen carrying capacity of his haemoglobin?

Answer:
Carbon monoxide forms an irreversible covalent bond with the haemoglobin to form a compound called carboxyhaemoglobin. Only the balance of the free unbounded haemoglobin is available for uptake of oxygen.
Therefore the arterial haemoglobin oxygen carrying capacity is $15/2 \times 1.34$ or 10.05 ml of oxygen per dl of blood when his haemoglobin is fully saturated.

Question:
What is the arterial oxygen content of Mr C?

Answer:
For Mr C, the amount of oxygen carried by the arterial blood

$\qquad$ = the amount of oxygen carried by the FREE haemoglobin

$\qquad$ + oxygen that is dissolved in the plasma.

$\qquad$ = (10.05 ml of oxygen) + 0.3 ml O_2/dl

$\qquad$ = 10.35 ml O_2/dl

Question:
What conclusion can you infer in the oxygen content of Mr B (in the previous question) and Mr C?

Answer:
A person with a haemoglobin level of 7.5 g/dl (Mr B) would have the same total oxygen content as a person with a haemoglobin level of 15 g/dl but who has 50 % of his haemoglobin in the carboxyhaemoglobin form. In patients with carbon monoxide poisoning, new haemoglobin has to be synthesis and release by the marrow in order for the total oxygen content to improve.

Question:
Back to Mr C, which of the conditions below, will be effective in improving his oxygen content:

(a) Hyperventilating
(b) Breathing 100 % oxygen at one atmospheric pressure.
(c) Breathing 100 % oxygen at three atmospheric pressure.

Answer:
Mr C's Haemoglobin is 50 % saturated (Free or unbounded with carbon monoxide). As such, the only way to increase his total oxygen content in the arterial blood is to increase the dissolved oxygen component.

(a) At best hyperventilating in room air will only increase his alveolar oxygen partial pressure to approach inspired oxygen at atmospheric pressure which is

$$(760 \, mm \, Hg - \text{partial pressure of water vapour}) \times 21/100$$
$$= (760 - 47) \times 21/100$$
$$= 713 \times 21/100$$
$$= 150 \, mm \, Hg$$

The dissolved component is then equal to 150 mm Hg $\times$ 0.003 ml O$_2$/dl = 0.45 ml O/dl.
There is only an increase of (0.45 − 0.3 ml) O$_2$/dl or 0.15 ml O$_2$/dl which is negligible.

(b) When breathing 100 % oxygen at one atmosphere, his inspires partial pressure of oxygen is equal to

$$(760 \, mm \, Hg - \text{partial pressure of water vapour}) \times 100/100$$
$$= 713 \, mm \, Hg$$

The dissolved component i s then equal to 713 $\times$ 0.003 ml O$_2$/dl = 2.1 ml O$_2$/dl.
There is only an increase of (2.1 − 0.3 ml) O$_2$/dl or 1.8 ml O$_2$/dl which is also fairly insignificant.

(c) At three atmospheric pressure and 100 % oxygen, his inspires partial pressure of oxygen is equal to

$$(3 \times 760 \, mm \, Hg - \text{partial pressure of water vapour}) \times 100/100$$
$$= 2233 \, mm \, Hg$$

The dissolved component is then equal to 2233 $\times$ 0.003 ml O$_2$/dl = 6.7 ml O$_2$/dl.
There is an increase of 6.7 − 0.3 ml O$_2$/dl or 6.4 ml O$_2$/dl in the dissolved blood which is a significant increased.

Therefore, when breathing 100 % oxygen at three atmospheric pressure, the amount of oxygen carried by the free haemoglobin and the amount which is dissolved in the plasma

$$= 10.35 + 6.7 \, \text{ml} \, O_2/\text{dl}$$
$$= 17 \, \text{ml} \, O_2/\text{dl}$$

This is a very significant increment and is one of the means used to treat carbon monoxide poisoning. It provides sufficient oxygen to the cells while awaiting new red blood cells to be formed and released from the marrow. This mode of treatment is termed hyperbaric oxygen therapy.

Note: According to the Hong Kong College of Anaesthesiologists the 'elimination ½ life of CO decreases from 250 min when breathing air to 59 min when breathing 100 % O_2 and 22 min when breathing 100 % O_2 at 2.2 atmospheres' (Hong Kong College of Anaesthesiologists 2004).

Question:
Comparing Mr B and Mr C, who do you think is in a more critical situation?

Answer:
Although, both their oxygen content is reduced to about the same level, Mr C is in a more critical situation. This is because carboxyhaemoglogin shifts the Oxy haemoglobin Dissociation Curve to the left, thereby increasing the free haemoglobin's affinity for the oxygen resulting in less available for the cells.

Carboxyhaemoglobin can also interfere with the cytochrome C oxidase at the mitochondrial level, hindering ATP transport. There is thus a functional hypoxic state at the mitochondrial level despite fairly adequate oxygen supply at the capillary level.

6.1.4 Theme: Oxygen Carrying Capacity 2

An adult male fireman was brought to the emergency department after being trapped in a burning building. He was diagnosed to be suffering from 100 % carbon monoxide poisoning where all his haemoglobin is saturated with carbon monoxide.

Question:
Is hyperbaric oxygen effective in this circumstance?

Answer:
This scenario demonstrates that oxygen carried in solution at hyperbaric condition is sufficient to maintain life while awaiting new haemoglobin to be restored by the marrow.

Oxygen is transported in the blood in two forms, either bound to haemoglobin (Hb) or dissolved in plasma.

The solubility of O_2 in plasma is low (0.003/mm Hg/100 ml) and in normobaric conditions is insignificant. Under normal circumstances, the greatest proportion of O_2 is transported bound to haemoglobin.

The amount of O$_2$ bound to haemoglobin can be calculated from the equation below:

$$\text{Hb bound } O_2(\text{mL}/100\,\text{ml}) = \text{Hb concentration (g)} \times \text{HbO}_2\text{capacity } (1.34\,\text{ml/g})$$
$$\times \text{ Hb saturation } (\%)$$

Assuming a Hb concentration of 15 g.
Saturations of 97 % for arterial blood when P$_a$O$_2$ = 100 mm Hg
Saturation of 75 % for venous blood when P$_v$O$_2$ = 40 mm Hg,

Values for arterial (CaO$_2$) and venous (CvO$_2$) oxygen contents are;

$$C_aO_2 - \text{Hb bound } O_2(\text{ml}/100\,\text{ml}) + O_2\text{in dissolved form}(0.003/\text{mm Hg}/100\,\text{ml})$$
$$= (15 \times 1.34 \times 0.97) + (0.003 \times 100)$$
$$= 19.5\,\text{ml}/100\,\text{ml} + 0.3\,\text{ml}/100\,\text{ml}$$
$$= 19.8\,\text{ml}/100\,\text{ml}.$$

$$C_aO_2 = \text{Hb bound } O_2(\text{ml}/100\,\text{ml}) + O_2\text{in dissolved form}(0.003/\text{mm Hg}/100\,\text{ml})$$
$$= (15 \times 1.34 \times 0.75) + (0.003 \times 40)$$
$$= 15.0\,\text{ml}/100\,\text{ml} + 0.12\,\text{ml}/100\,\text{ml}$$
$$= 15.1\,\text{ml}/100\,\text{ml}.$$

The relative quantitative importance of O$_2$ transported by haemoglobin as compared with the dissolved component is demonstrated.

It is also seen that under normal normobaric conditions, about 5 mL/100 ml of O$_2$ (CaO$_2$ − C$_v$O$_2$) is extracted from arterial blood to meet the body's metabolic requirement.

Since haemoglobin is almost fully saturated (97 %) when breathing room air at one atmospheric pressure, there is little potential for increasing O$_2$ transport by the haemoglobin component by increasing the partial pressure of oxygen using hyperbaric oxygenation. In contrast, the dissolved O$_2$ component increases linearly with increasing partial pressure of oxygen. This component is significant in hyperbaric conditions with a high F$_I$O$_2$.

For example,

When breathing air at three atmospheres, the dissolved O$_2$ component is 1.3 ml/100 ml
When breathing 100 % O$_2$ at three atmospheres, the dissolved O$_2$ results in 6.7 ml/100 ml.

Therefore, breathing 100 % O$_2$ at three atmospheres results in a dissolved fraction (6.7 ml/100 ml) which is sufficient to meet the body's metabolic requirement at rest (5 ml/100 ml) despite the absence of functional haemoglobin. This forms the basis of using hyperbaric oxygen in conditions where oxygen delivery is compromised by absence of functional haemoglobin such as in carbon monoxide poisoning.

Theoretically, if human being were to colonies a planet where the atmospheric pressure and partial pressure of atmospheric oxygen is the same as in the hyperbaric chamber as above, then human being may not need haemoglobin to survive in that planet.

Another way to appreciate this learning scenario is to recognize that a patient bleeding to a level of ZERO haemoglobin can be kept alive in a hyperbaric chamber while awaiting blood transfusion.

6.1.5   Theme: Oxygen Cascade and Alveolar Gas Equation

A healthy person is breathing room air. He is then allowed to breathe 100 % oxygen.

Question:
Calculate the partial pressure of oxygen in both instances in the

(a)   inspired air
(b)   tracheal
(c)   alveolar

Answer:

(a)   Air contains 21 % oxygen. Therefore when breathing room air, the inspires partial pressure of oxygen is

$$(21/100) \times 760 \, \text{mm Hg} \approx 160 \, \text{mm Hg}$$

When breathing 100 percent oxygen, the inspired partial pressure of oxygen is

$$(100/100) \times 760 = 760 \, \text{mm Hg}.$$

(b)   When Air goes through our upper airways, it becomes humidified and becomes fully saturated with a water vapour pressure of 47 mm Hg. Therefore when breathing room air, the partial pressure of oxygen in the tracheal is

$$(21/100) \times (760 - 47 \, \text{mm Hg}) \approx 150 \, \text{mm Hg}$$

When breathing 100 % oxygen, the partial pressure of oxygen in the tracheal is

$$(100/100) \times (760 - 47 \, \text{mm Hg}) = 713 \, \text{mm Hg}$$

(c)   The partial pressure of oxygen in the alveolar is calculated using the **alveolar gas equation.**

$$P_A O_2 = P_I O_2 - P_A CO_2 / R$$

Assuming just before mixing, arterial CO_2 is equals alveolar CO_2, then

$$P_AO_2 = P_IO_2 - P_aCO_2/R$$

where P_AO_2 = Partial pressure of oxygen in the alveolar gas, P_IO_2 = Partial pressure of oxygen in the inspired gas, P_aCO_2 = Partial pressure of carbon dioxide in the alveolar gas. P_aCO_2 = Partial pressure of carbon dioxide in the arterial blood. R = Respiratory Quotient

Therefore, the partial pressure of oxygen in the alveolar when breathing room air is

$$P_AO_2 = 150 - P_aCO_2/0.8$$

$P_AO_2 = 150 - 40/0.8$ (assuming $P_aCO_2 = 40$ mm Hg, obtained from arterial blood gas analysis, and R $-$ 0.8)

$$P_AO_2 = 100 \, mm \, Hg.$$

Similarly, when breathing 100 % oxygen, the partial pressure of oxygen in the alveolar is

$$P_AO_2 = 713 - 40/0.8$$
$$P_AO_2 = 663 \, mm \, Hg.$$

6.1.6 Theme: Alveolar-Arterial Oxygen Gradient (A-a) O_2: 1

A 70 year old man who is known to have chronic obstructive airway disease arrived at the emergency department complaining of shortness of breath. An arterial blood gas study was performed when he was breathing room air.

The results were

$$PaO_2 = 50 \, mm \, Hg, \quad PaCO_2 = 64 \, mm \, Hg$$

Question:
How do you identify the aetiology of this patient's hypoxemia?

Answer:
To diagnose the aetiology of hypoxemia, we use the **Alveolar–arterial oxygen gradient** (A-a) O_2 gradient, which is a measure of the difference between the **alveolar** concentration (A) of oxygen and the **arterial** (a) concentration of oxygen.

Firstly calculate the P_AO_2, from the Alveolar Gas Equation

$$P_AO_2 = P_IO_2 - P_aCO_2/R$$
$$P_AO_2 = (21/100) \times (760 - 47 \, mm \, Hg) - 64/0.8$$
$$P_AO_2 = 150 - 80$$
$$= 70 \, mm \, Hg.$$

Therefore,

$$(A\text{-}a)O_2 = 70 - 50 mm \, Hg$$
$$= 20 \, mm \, Hg$$

The normal **Alveolar–arterial oxygen gradient** is less than 15 mm Hg, (+3 mm Hg each decade over 30 years). In this instance there is an elevated gradient.

An abnormally increased (A–a) O_2 gradient suggests a defect in each of the following

(I)   Diffusion defect
(II) V/Q (ventilation/perfusion ratio) mismatch
(III) right-to-left shunt.

The most probable cause of hypoxemia in this patient would be either a diffusion impairment or a V/Q mismatch as a result of chronic obstructive airway disease. The hypercarbia is indicative of hypoventilation.

6.1.7 Theme: Alveolar-Arterial Oxygen Gradient (A-a) O_2: 2

A patient with shortness of breath was given supplemental oxygen. His arterial blood gas on 30 % oxygen was

$$PaO_2 = 102 \, mm \, Hg, PaCO_2 = 40 \, mm \, Hg$$

Base on the arterial blood gas result, a junior doctor may conclude that the patient has no lung pathology since the partial pressures of the oxygen and carbon dioxide appears 'normal'. There is no hypoxemia in this case.

Question:
What further 'test' can you perform to indicate otherwise?

Answer:
One must have a high suspicion of index when arterial blood gas results are 'normal' especially when patients are on supplemental oxygen. A simple calculation to determine the **Alveolar-arterial oxygen gradient (A-a) O_2** will indicate whether the patient has any underlying pathology.

In this case, using the Alveolar gas equation

$$P_AO_2 = P_IO_2 - P_aCO_2/R$$
$$P_AO_2 = (30/100) \times (760 - 47 \, mm \, Hg) - 40/0.8$$
$$P_AO_2 = 214 - 50 \, mm \, Hg$$
$$= 160 \, mm \, Hg.$$

Therefore,

$$(A\text{-}a)O_2 = 164 - 102$$
$$= 62 \, mm \, Hg$$

The (A-a) O_2 is grossly elevated.

Although the P_aO_2 may appear 'normal', we have to take into consideration the supplemental oxygen therapy otherwise we will miss, a large (A-a) O_2 gradient. Therefore we must always maintain a high index of suspicion whenever patients are on oxygen therapy and calculate out the (A-a) O_2 gradient.

An abnormally increased A-a gradient suggests a defect in diffusion, V/Q (ventilation/perfusion ratio) mismatch, or right-to-left shunt.

6.1.8 Theme: Alveolar-Arterial Oxygen Gradient (A-a) O_2: 3

A patient was noted to be hyperventilating in the emergency department. An arterial blood gas analysis showed

$$PaO_2 = 80 \, mm \, Hg, \, PaCO_2 = 20 \, mm \, Hg$$

Question:
Interpret this arterial blood gas.

Answer:
A person who is hyperventilating should be inhaling in more oxygen into the lungs. Unless we perform a calculation to determine the (A-a) oxygen gradient, we may miss out on underlying lung pathology especially if patients are on oxygen therapy.

For this patient, applying the Alveolar gas equation

$$P_AO_2 = P_IO_2 - P_aCO_2/R$$
$$P_AO_2 = (21/100) \times (760 - 47 \, mm \, Hg) - 20/0.8$$
$$P_AO_2 = 150 - 25 \, mm \, Hg$$
$$= 125 \, mm \, Hg.$$

Therefore,

$$(A\text{-}a) \, O_2 = 125 - 80$$
$$= 45 \, mm \, Hg$$

The (A-a) O_2 is grossly elevated.

An abnormally increased A-a gradient suggests a defect in diffusion, V/Q (ventilation/perfusion ratio) mismatch, or right-to-left shunt.

Chapter 7
Regulation of Respiration

Abstract This chapter introduces the readers to the applications of the fundamentals of physiology to provide an understanding of the pathophysiology of some common respiratory illnesses. Students should always strive to comprehend the pathophysiology of diseases rather than try to commit them to memory.

Keywords Cor pulmonale · Hypoxic pulmonary vasoconstriction · Hypoxic drive · Haldene effect · Hypercarbic drive · Pulse oximeter

7.1 Introduction

The role of the hypoxic pulmonary vasoconstriction (HPV) in the pathogenesis of chronic obstructive pulmonary diseases and the paradoxical danger of oxygen in its therapy is afforded greater emphasise in this chapter.

Carbon monoxide is revisited by from the pathogenesis point of view. A short reference of pulse oximetry is included to demonstrate its usefulness or lack of it in the monitoring of carbon monoxide poisoning.

7.1.1 Theme: Cor Pulmonale

Mr X, a 60 year old chronic heavy smoker presented at the emergency department with shortness of breath and ankles swelling. On examination he has signs consistent with cor pulmonale.

Question:
What do you understand by the term 'Cor Pulmonale'?

© The Author(s) 2015   77
K.K. Mah and H.M. Cheng, *Learning and Teaching Tools for Basic
and Clinical Respiratory Physiology*, SpringerBriefs in Physiology,
DOI 10.1007/978-3-319-20526-7_7

Answer:
It is the alteration of the structure and function of the right ventricle resulting from diseases affecting the function or structure of the lungs. The end result is right heart failure.
Question:
What is the pathogenesis of Cor Pulmonale in this patient?

Answer:
This patient has chronic obstructive pulmonary disease (COPD) which leads to pulmonary hypertension which in turn results in right ventricular enlargement (hypertrophy or dilatation). If left undiagnosed or untreated, it will lead to right ventricular failure. Pulmonary hypertension is often the underlying pathological mechanism for right ventricular hypertrophy or failure in cor pulmonale.

Question:
How does COPD lead to pulmonary hypertension?

Answer:
Classically, pulmonary hypertension in COPD has been considered to be the result of hypoxic pulmonary vasoconstriction (HPV). Recently, it has been recognized that hyperinflation, inflammation and immune responses in COPD are also important contributory factors in the pathogenesis of pulmonary hypertension.

Question:
Elaborate further how HPV can lead to pulmonary hypertension and subsequently to cor pulmonale.

Answer:
Hypoxia causes vasoconstriction in small pulmonary arteries in the pulmonary circulation BUT vasodilatation in the systemic circulation.

Hypoxic pulmonary vasoconstriction (HPV), also known as the von Euler-Liljestrand mechanism, is a physiological adaptive response to alveolar hypoxia. This is a physiological protective mechanism which re-distributes pulmonary capillary blood flow from alveoli with low oxygen partial pressure to those with high oxygen partial pressure. As a result local lung perfusion is then matched to areas of good ventilation. Impairment of this compensatory mechanism may result in hypoxaemia.

Under conditions of chronic hypoxia as in this case, generalised vasoconstriction of the pulmonary vasculature in concert with hypoxia-induced vascular remodelling leads to pulmonary hypertension. Although the principle of HPV was acknowledged decades ago, its exact pathway is not fully understood. Many vaso-active substances have been implicated as possible mediators or modulators of HPV.

Question:
What are other signs of cor pulmonale which may be seen in this patient?

Answer:
The signs of cor pulmonale are attributed to right heart failure. In right heart failure, the cardiac output of the right ventricle is impaired leading to elevated jugular venous pressure, liver congestion, jaundice, ascites and peripheral oedema.

Question:
Would you expect to see pulmonary edema in this case?

Answer:
There is no pulmonary edema in right heart failure which is a feature of left heart failure.

Note: This is a favourite question during a teaching ward round. This question tests the ability of students to understand the differences in the pathogenesis of right heart as compared to left heart failure.

7.1.2 Theme: Chronic Obstructive Pulmonary Disease (COPD)

Mr X, a 70 year old patient with chronic obstructive pulmonary disease required admission because of chest infection. His blood gas result showed PaO_2 of 60 mm Hg and $PaCO_2$ of 55 mm Hg when breathing air. When supplementary high flow oxygen was given, his respiratory effort deteriorated and a new blood gas results revealed PaO_2 of 50 mm Hg and a $PaCO_2$ of 60 mm Hg.

Question:

1. Describe the mechanisms involved in causing an increased of $PaCO_2$ and a decreased of PaO_2 when supplementary **high concentration of** oxygen is given to this patient.

Answer:
The mechanisms whereby the CO_2 in the blood may increase further in this COPD patients when given high inspired oxygen, include

(a) Loss of hypoxic drive
(b) Ventilation/perfusion mismatch
(c) Haldene effect.

(a) Loss of hypoxic drive resulting in hypoventilation.

In a normal healthy person, the dominant ventilatory drive is the hypercarbic drive which is attributable to carbon dioxide acting on the central chemo-receptors in the medulla. Classically, it was believed that in patients with chronically elevated CO_2 such as in COPD patients, their dominant central respiratory drive is desensitised. This patient buffers the increasing respiratory acidosis by a compensatory metabolic alkalosis. The retained bicarbonate of metabolic alkalosis will in turn reduce the H^+ ion stimulation of the central respiratory centre. This desensitization results in further hypoventilation, which in turn results in hypoxemia and more severe hypercarbia. The body will then begin to rely more on the hypoxic drive and less on the hypercarbic drive to maintain ventilation.

As the name suggest, the hypoxic drive acts in situation of hypoxemia to drive ventilation. Hypoxia will stimulate the peripheral chemo-receptors (the aortic and carotid bodies located in the aortic arch and at the junction of the carotid arteries respectively).

When supplemental high concentration of oxygen is delivered to this patient, the resultant initial elevated PaO_2 will suppress the peripheral chemo-receptors. This will then lead to a further deterioration of respiratory effort or hypoventilation. (Remember, the dominant hypercarbic drive has been previously desensitized by elevated $PaCO_2$)

This is further explained if one looks at the Oxygen Ventilatory Response Curve below. The increase ventilatory response to hypoxemia only really kicks in at around PaO_2 of 60 mm Hg. Supplementary oxygen especially at high concentration will initially increase PaO_2 beyond 60 mm Hg (to the right) and thus depresses the ventilatory response to hypoxia. At the horizontal part of the curve beyond PaO_2 of 60 mm Hg, the ventilator response to PaO_2 is fairly constant.

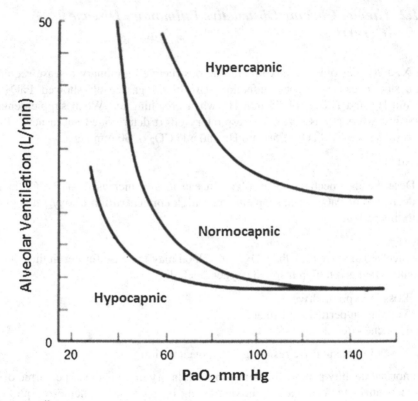

Oxygen ventilatory response curve

Patients identified as, most at risk, are those who are chronically hypoxic with moderate to severe hypercarbia.

Note: We can consider the hypoxic drive as a spare ventilatory driver, 'kicking in' when the hypercarbic drive is impaired or desensitised. This 'kicking in' occurs when the PaO_2 has dropped to 60 mm Hg.

(b) Ventilation/perfusion mismatch

In a healthy individual, alveolar ventilation and perfusion are fairly matched physiologically. Two extremes of ventilation-perfusion (V/Q) mismatch however may occur, a shunt (no ventilation of an alveolus but adequate perfusion) or a dead space (adequate ventilation but no perfusion).

The body has adaptive mechanisms to optimise the V/Q ratio. When alveolar oxygen tension decreases (for example, hypoxia due to bronchoconstriction), local mediators induce vasoconstriction of local pulmonary capillaries supporting this specific alveolus, counteracting any potential shunting. This mechanism is known as hypoxic pulmonary vasoconstriction (HPV).The strongest mediator for hypoxic pulmonary vasoconstriction is low alveolar pO_2 (partial pressure of oxygen).

Therefore, unknowingly providing a high concentration of inspired O_2 (FiO_2) will increase O_2 tension in poorly ventilated alveoli and therefore **inhibiting** the protective hypoxic pulmonary vasoconstriction. As a result, the poorly ventilated alveoli are now needlessly well perfused instead of vaso-constricted, leading to an increase in V/Q mismatch. This results in further low PaO_2 and increased $PaCO_2$.

In summary, in patients with COPD, hypoxic pulmonary vasoconstriction is the most efficient and effective adaptive mechanism to adjust the V/Q ratios to improve gas exchange. This physiological mechanism is inhibited by oxygen therapy and accounts for the largest increase of oxygen-induced hypercapnia and hypoxemia.

(c) Haldene effect

Amine groups in haemoglobin, combine readily with CO_2 to form carbamino compounds. Also, haemoglobin can exist in two forms: an oxygenated and a deoxygenated form. It is the deoxygenated haemoglobin which has a higher affinity to bind CO_2 to form carbamino compounds. Put in another way, oxygen causes a decrease in the affinity of the haemoglobin for carbon dioxide and induces a rightward shift of the CO_2 dissociation curve. This is known as the Haldane effect.

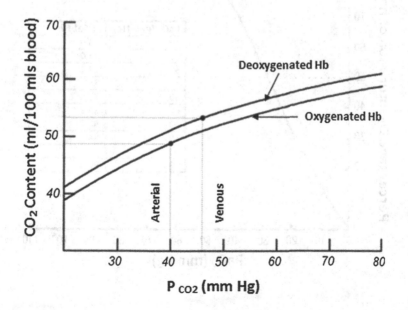

A rightward shift in the CO_2 dissociation curve will decrease CO_2 content of the haemoglobin and increase $PaCO_2$, which in the normal person, is removed through hyperventilation, thereby normalizing $PaCO_2$.

In summary, in patients with severe COPD, who are unable to increase minute ventilation due to a desensitised central respiratory chemoreceptors, administration of O_2, will increase $PaCO_2$ due to the Haldane effect. $PaCO_2$ will remains high in the blood as there will be no increased in ventilation for carbon dioxide homeostasis.

Question:

2. It is usually recommended that the haemoglobin saturation of COPD patients be maintained at about 90 % when they are receiving oxygen therapy. What is the rationale of keeping the oxygen saturation at this level?

Answer:

Administration of high oxygen concentrations which is associated with additional hypoxemia and hypercarbia as described above should be avoided.

A titrated oxygen therapy to achieve saturations of 88–92 % is recommended to avoid hypoxemia and reduce the risk of oxygen-induced hypercapnia.

From the oxygen dissociation curve below, haemoglobin saturation at 90 % would be equivalent to PaO_2 of 60 mm Hg.

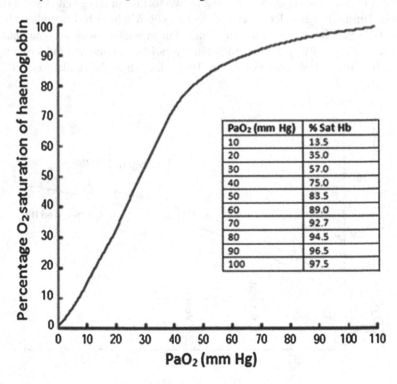

PaO₂ (mm Hg)	% Sat Hb
10	13.5
20	35.0
30	57.0
40	75.0
50	83.5
60	89.0
70	92.7
80	94.5
90	96.5
100	97.5

As discussed in (a), maintaining the PaO_2 around a haemoglobin saturation of 90 % which is equivalent to a PaO_2 of 60 mm Hg ensures a balance of retaining the effectiveness of the hypoxic drive and preventing hypoxemia (Oxygen Ventilatory Response Curve). Too high a PaO_2 or haemoglobin saturation beyond 92 % will inhibit the hypoxic drive.

7.1.3 Theme: Carbon Monoxide Poisoning

A couple was found in the morning semi-conscious inside a tent in a park. Apparently the couple was burning charcoal to heat up the tent. The ambulance arrived and they were rushed to the hospital for treatment.

Question:

(a) What do you think is the most likely cause of the altered consciousness?
(b) Describe its pathogenesis.

Answer:

(a) The couple probably succumbed to carbon monoxide poisoning which is derived from incomplete combustion of the charcoal or other fuels such as gas, oil or wood. This problem is further compounded by the carbon monoxide being trapped in a closed environment like inside a tent in this case.
(b) After breathing in carbon monoxide, it enters your bloodstream and mixes with haemoglobin to form carboxyhaemoglobin (COHb). The affinity between haemoglobin and carbon monoxide is approximately 200 times stronger than the affinity between haemoglobin and oxygen, so haemoglobin binds preferably to carbon monoxide. This decreases the oxygen-carrying capacity of the blood and hinders the transport and delivery of oxygen to the cells where they are needed for oxidative metabolism and the production of adenosine triphosphate. When this happens, tissue hypoxia, inadequate cellular oxygenation, impaired and organ function and eventually death will occur.

An alternative theory of carbon monoxide poisoning relates to haemoglobin being a tetramer with four oxygen binding sites. Binding of carbon monoxide at one of these sites increases the oxygen affinity at the remaining three sites. What happens then is that the haemoglobin molecule will hang on to the oxygen molecules and not release them to the tissues. The end result is tissue hypoxia. This can explain why the oxygen dissociation curve is shifted to the left in the presence of carbon monoxide.

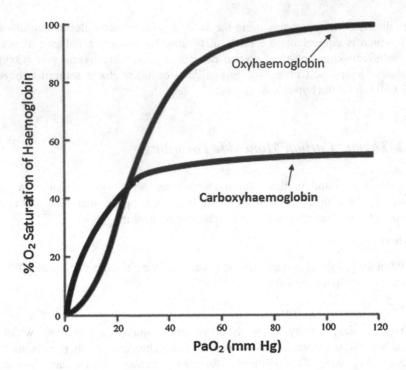

Question:
Will the couple be aware of the impending danger of carbon dioxide poisoning?

Answer:
Carbon monoxide poisoning is difficult to detect because it has no smell, no taste, colourless and is does not irritate the senses. This means you can inhale carbon monoxide without realizing it. Because of its innocuous properties, carbon monoxide is sometimes referred to as the "silent killer". The person literary slips into deep sleep and never wake up.

It is because of this 'painless' and 'peaceful' death, carbon monoxide poisoning is a widely used method of committing suicide whereby the person attaches a hose to the car exhaust system and insert the other end into the passenger compartment with the windows wind down and engine running.

Question:
Are there any signs to suggest that the patient is suffering from carbon monoxide poisoning?

Answer:
Signs of poisoning include: cherry red skin and mucous membranes as a result of bright red carboxyhaemoglobin.

In the emergency department the patients were examined and stabilised. He was monitored with a pulse oximeter which showed an oxygen saturation of 97 % and arterial blood gases were obtained.

Question:
Is pulse oximetry reliable in this case?

Answer:
Pulse oximeter is not reliable in the diagnosis of carbon monoxide poisoning. Pulse oximeters utilises 2 wavelengths of light to detect oxygenated and deoxygenated haemoglobin in the blood. Unfortunately light absorbance at one of the wave length (660 nm) is very similar for both oxyhaemoglobin and carboxyhaemoglobin and as such the pulse oximeter will be unable to differentiate between the two forms of haemoglobin. As a result, carboxyhaemoglobin will be measured as oxygenated haemoglobin and such patients can have a spuriously normal oxygen saturation level. Pulse oximetry may overestimate oxyhaemoglobin by the amount of carboxyhaemoglobin present. In other words, a person with 20 % carboxyhaemoglobin, plus 75 % oxygenated haemoglobin in his blood, can have a pulse oximeter reading of 95 % saturation.
To detect COHb, a CO-oximeter is required.

Question:
What would you expect the partial pressure of oxygen to be in this patient?

Answer:
Carbon monoxide poisoning causes anaemic hypoxia. In anemic hypoxia, the blood oxygen partial pressure is normal. What is lacking is normal haemoglobin for oxygen carriage.

Chapter 8
Respiratory Physiology in Hostile Environment

Abstract Partial pressures and concentration of gas are emphasied with respect to ambient atmospheric pressure in this chapter. This brings in the various laws such as by Dalton, Henry, Boyles and Reynolds number. The various scenarios will be able to bring 'life' into these otherwise 'mundane' laws.

Keywords High altitude · Sudden decompression · Henry's law · Reynolds number · Oxygen toxicity · Nitrogen narcosis · Decompression sickness

8.1 Introduction

This concluding chapter takes us on a journey to the deep sea and also to the final frontier that is, outer space. The journey begins in Mt. Everest and the outer space and ends in the deep blue ocean. The atmospheric pressures are extreme at both locations and their effects on respiratory physiology can be daunting. Each environment has its own unique physiological problems and we believed that if the readers were to go through all the scenarios provided, clinical and non clinical, they will be able to have a deeper understanding of the fundamentals of respiratory physiology in an otherwise hostile environment.

8.1.1 Theme: High Altitude Respiratory Physiology

A mountain climber on a Mount Everest expedition initially climbed from sea level to the Everest base camp which is at a height of 5000 m. He then attempted to ascend to the summit which is at a height of 8848 m. He experienced shortness of breath from 5000 m and required supplemental oxygen.

© The Author(s) 2015
K.K. Mah and H.M. Cheng, *Learning and Teaching Tools for Basic and Clinical Respiratory Physiology*, SpringerBriefs in Physiology,
DOI 10.1007/978-3-319-20526-7_8

87

Question:

What is the concentration of oxygen at sea level, the base camp and the summit?

Answer:

The concentration of oxygen at all the different levels is the same, which is 21 %. This is true up to a level of about 100 km above sea level.

Question:

Why did the mountain climber experienced shortness of breath at 5000 m and above?

Answer:

Although the concentration of oxygen is the same from sea level up to an altitude of 100 km, it is the lower partial pressure of oxygen at high altitude that causes hypoxemia and subsequently shortness of breath.

At the base camp which is at an altitude of 5000 m, the atmospheric pressure is about half that at sea level. On the summit, the atmospheric pressure is about one third that at sea level.

The partial pressures of oxygen (PO_2) = FO_2 × Barometric pressure at the various sites:

(a) Sea level = 0.21×760 mm Hg = 160 mm Hg
(b) Base camp (5000 m) = $0.12 \times 380 = 80$ mm Hg
(c) Summit (8848 m) = $0.12 \times 253 = 53$ mm Hg

Question:

What is the estimated arterial oxygen partial pressure of oxygen at

(a) sea level
(b) base camp and
(c) the summit?

This can be derived from the Alveolar Gas Equation, where

$$P_AO_2 = P_IO_2 - P_aCO_2/R$$
$$P_AO_2 = [F_IO_2(P_{atm} - P_{H2O}] - P_aCO_2/R$$

Since an estimated result is required, we can make some assumptions: $P_aCO_2 = 40$ mm Hg at sea level and 30 mm Hg at higher altitude (As a result of hyperventilation), $P_{H2O} = 47$ mm Hg and R = 0.8 and $\mathbf{P_aO_2 \propto P_AO}$

Then the estimated P_aO_2 at

(a) Sea level = $(21/100) \times (760 - 47$ mm Hg$) - 40/0.8$

   = 100 mm Hg

(b) Base camp = $(21/100) \times (380 - 47$ mm Hg$) - 30/0.8$

   = 32 mm Hg

(c) Summit = $(21/100) \times (253 - 47$ mm Hg$) - 30/0.8$

   = 6 mm Hg

From the above, it can be deduced that, it would be near impossible to survive at the summit without oxygen supplementation. In fact, in mountaineering, there is an altitude above a certain point where the amount of atmospheric oxygen is insufficient to sustain human life. This is known as the 'Death Zone'. This zone is generally at 8000 m.

8.1.2 Theme: Sudden Decompression in a Jumbo Jet

A commercial passenger jet suffered a sudden loss of cabin pressure due to a malfunctioned door while cruising at an altitude of 33,000 ft. or 10,300 m. At this altitude the outside atmospheric pressure is about 200 mm Hg.

Question

(a) What is the partial pressure of oxygen outside the plane at this altitude?
(b) What is the partial pressure of oxygen inside the plane prior to the sudden decompression at this altitude?
(c) What would be the estimated arterial partial pressure in a healthy adult?
(d) What is the effect of this sudden loss of cabin pressure on the passengers?

Answer:

(a) At an altitude of 10,300 m, the partial pressure of oxygen is equal to:

Atmospheric pressure at 10,300 m $\times$ O_2 concentration

$$= 200 \text{ mm Hg} \times 0.21$$

$$= 42 \text{ mm Hg}$$

This is less than at the summit of Mt. Everest (8848 m) which has an oxygen partial pressure of 53 mm Hg.
(b) The cabin pressure in commercial airplanes are pressurised to 600 mm Hg which corresponds to an altitude of 2000 m.
At this height, the partial pressure of oxygen inside the airplane is

$$600 \text{ mm Hg} \times 0.21$$

$$= 126 \text{ mm Hg}.$$

This is fairly comfortable for most healthy person except those with cardio-respiratory problem.
(c) The estimated partial pressure of oxygen in a healthy passenger can again be deduced from the alveolar gas equation:

$$P_AO_2 = P_IO_2 - P_aCO_2/ R$$

$$P_AO_2 = [F_IO_2(P_{atm} - P_{H2O}] - P_aCO_2/ R$$

Since an estimated result is required, we can make some assumptions: $P_aCO_2 = 40$ mm Hg and $R = 0.8$ and $P_aO_2 \propto P_AO$. The partial pressure of water vapour(P_{H2O}) is assumed to be 20 mm Hg as the humidity is kept lower in the airplane.
Therefore

$$P_aO_2 = [0.21(600 - 20)] - 40/0.8$$
$$= 122 - 50 \text{ mm Hg}$$
$$= 72 \text{ mm Hg}$$

With a P_aO_2 of 72 mm Hg, his arterial blood would be almost 93 % saturated based on the oxygen dissociation curve.

(d) A sudden decompression will cause a sudden drop in P_AO_2 resulting in severe hypoxemia and loss of consciousness.

To overcome this oxygen mask must be used to maintained oxygen delivery to the brain.

Pilots are trained to descend rapidly to a lower altitude where the atmospheric pressure is higher.

Failure to rapidly descend when the supplemental oxygen is exhausted will lead to death due to hypoxemia and severe hypothermia.

8.1.3 Theme: Sudden Decompression in Space

A space station orbiting Earth at 25 km in the stratosphere is hit by a meteorite causing sudden massive decompression.

Question:
What would be the effect of this sudden decompression on the occupants of the space station?

Answer:
If the astronauts are not in their pressure suits, they will die from hypoxemia, hypothermia and massive air embolism. In fact the blood of the astronauts will boil.

The boiling point of a substance is the temperature at which the vapour pressure of the liquid is equal to the pressure surrounding the liquid. In other words, the boiling point of a liquid varies depending upon the surrounding environmental pressure.

The water vapour pressure in the alveoli at room temperature is 47 mm Hg. The barometric pressure at 25 km in space is almost 0 mm Hg.

Therefore by definition, the water in the alveoli and the whole body will boil when the body is exposed to a pressure less than or equal to 47 mm Hg. Theoretically the body volume of a human will increase three times by air expansion in the blood and tissue on sudden decompression in space (Boyle's Law).

8.1.4 Theme: Snorkeling

A 50 kg young man goes snorkeling.

Question:
Assuming his tidal volume is 600 ml and the radius of his snorkeling tube is 1 cm, what is the theoretical deepest depth he can descend to?

Answer:

This is a dead space problem. When the diver's respiratory dead space and the snorkel dead space are equal to the alveolar tidal volume, no gas exchange will occur even if he continues breathing.

Assuming his dead space is represented by his anatomical dead space (omitting alveolar dead space), his anatomical dead space will be 110 ml (Anatomical dead space is 2.2 ml/kg)

Effective alveolar ventilation would cease when the snorkel dead space is

= tidal volume minus anatomical dead space

= (600 − 110) ml

= 490 ml

Each centimetre length of the snorkel has a dead space volume of

$$\pi r^2 \times 1\,cm = 3.14 \times 1 \times 1 \times 1$$
$$= 3.14\,cm^3$$

Therefore the theoretical depth when no effective ventilation occurs is:

490/3.14 = 156 cm.

Question:

In reality, what is the usual depth that a diver with a snorkel can descend?

Answer:

Most snorkels have a tube of about 30 cm long and with an internal diameter of about 2 cm. The optimum length of the snorkel tube is about 35 cm. A longer tube would hinder ventilation when snorkeling at greater depth, since it would subject the lungs to a greater external surrounding water pressure. The high-density water all around will press inwards from all sides. This means that if you were to try to breathe through a long snorkel at great depth, your chest would need to expand outwards against the weight of the water to inflate the lungs during inhalation. This excessive extra thoracic pressure would therefore limit the ability of the thorax to expand on inhalation.

As such, in reality, the diver would not have been able to ventilate at the calculated depth of 156 cm as in the example above.

Another problem with dead space is the buildup of carbon dioxide in the blood causing hypercapnia. This problem is greater if the dead space is large and the diver breaths with small tidal volumes.

8.1.5 Theme: Diving

A 30 year old scuba diver is working at a depth of 30 m in the sea.

Question:
Calculate the ambient pressure the diver is subjected to at this depth.

Answer:
The ambient pressure in water is a combination of the hydrostatic pressure due to the weight of the water column and the atmospheric pressure on the surface. This increases approximately linearly with depth. Each 10 m (33 ft.) of depth adds one bar or 101.3 kPa or 760 mm Hg to the ambient pressure.

Therefore at a depth of 30 m, the ambient pressure is equivalent to 4 bars. (Pressure due to a water column of 30 m and the atmospheric pressure of 1 bar at the surface.)

Question:
What does SCUBA means and what is the common composition of gas in recreational scuba diving?

Answer:
Scuba is an acronym for self-contained underwater breathing apparatus. Most recreational divers use scuba tanks filled with simple compressed air which are filtered and dehumidified. A professional diver who may dive at great depth and stays underwater for a prolong period will requires the use of specialised gas mixtures in their scuba tanks.

Question:
Calculate the partial pressure of oxygen in the alveolar air of the diver at the surface and at a depth of 30 m.

Answer:
The partial pressure of oxygen in the alveolar air is calculated using the Alveolar Gas Equation

$$P_AO_2 = P_IO_2 - P_ACO_2/R$$

Assuming just before mixing, arterial CO_2 equals alveolar CO_2, then

$$P_AO_2 = P_IO_2 - P_aCO_2/R$$

where P_AO_2 = Partial pressure of oxygen in the alveolar gas, P_IO_2 = Partial pressure of oxygen in the inspired gas, P_ACO_2 = Partial pressure of carbon dioxide in the alveolar gas. P_aCO_2 = Partial pressure of carbon dioxide in the arterial blood. R = Respiratory Quotient = 0.8

At the surface, the

$$
\begin{aligned}
P_AO_2 &= P_IO_2 - P_aCO_2/R \\
&= [F_IO_2(P_{ambient} - P_{H2O})] - P_aCO_2/R \\
&= (21/100) \times (760 - 47 \text{ mm Hg}) - 40/0.8 \\
&= 100 \text{ mm Hg}
\end{aligned}
$$

At a depth of 30 m, the

$$P_AO_2 = P_IO_2 - P_aCO_2/R$$
$$= [F_IO_2(P_{ambient} - P_{H2O})] - P_aCO_2/R$$
$$= (21/100) \times [(760 \times 4) - 47\,mm\,Hg] - 40/0.8$$
$$= 578\,mm\,Hg$$

(Assuming $P_aCO_2/R = 40/0.8$)

8.1.6 Theme: Oxygen Toxicity

The scuba diver above then descends to 60 m. His companion noted that he is behaving abnormally and appears disorientated.

Question:
What are the probable causes of this behaviour and how can this be minimised?

Answer:
At a depth of 60 m, his

$$P_AO_2 = P_IO_2 - P_aCO_2/R$$
$$= [F_IO_2(P_{ambient} - P_{H2O})] - P_aCO_2/R$$
$$= (21/100) \times [(760 \times 7) - 47\,mm\,Hg] - 40/0.8$$
$$= 1107 - 50\,mm\,Hg$$
$$= 1057\,mm\,Hg$$

(Assuming $P_aCO_2/R = 40/0.8$)

At this depth, he is probably suffering from acute oxygen toxicity. Central nervous system toxicity is caused by acute exposure to high partial pressures of oxygen at greater than atmospheric pressure. At a PiO_2 greater than 1200 mm Hg, acute oxygen toxicity can occur within minutes causing disorientation and convulsions with little or no warning. Prolonged breathing of a gas with a PiO_2 greater than 450 mm Hg can lead to pulmonary toxicity and eventually irreversible pulmonary fibrosis, but this takes many hours or days.

To avoid such fatal consequences of oxygen toxicity, divers should ensure their Pio_2 is below 1200 mm Hg.

8.1.7 Theme: Dalton's Law, Reynold's Number

A professional diver planned to go for a deep dive to a depth of 200 metres. He selected a scuba tank containing a mixture of 99 % helium and 1 % oxygen instead of the normal gas mixture of compressed air.

Question:

(a) What is Dalton's law and Reynolds number (Re) and their applications to respiratory physiology in diving?
(b) Explain the rationale of his choice for this gas mixture instead of compressed air in his scuba tank.
(c) What other advantages are there in using helium as a carrier gas?

Answer

(a) Dalton's Law states that in a mixture of gases, the partial pressure of the individual gas is proportional to their concentration in the mixture. Partial pressure is important in determining the effects of the component gases on divers. In diving in particular it is a useful measure to determine limits for avoiding oxygen toxicity and nitrogen narcosis.

Reynolds number (Re) is a dimensionless quantity that is used to help predict flow patterns. Laminar flow occurs at low Reynolds numbers (Re < 1200) and turbulent flow occurs at high Reynolds numbers (Re > 2000). Mathematically it is expressed as

$$Re = \rho v L / \mu$$

where ρ = density, v = velocity, L = length of the tube, μ = viscosity

(b) At a depth of 200 m, the ambient pressure is equivalent to 21 atm. (Every 10 m of sea water = 1 atm of pressure. Therefore the ambient pressure a diver experiences is equivalent to the water pressure plus 1 atm pressure).

At this depth, where the ambient pressure is 21 atm or (21 × 760) mm Hg,

I.   The partial pressure of oxygen (21 %) and nitrogen (79 %) in the compressed air in the scuba tank would be (21 × 760 × 0.21 mm Hg) and (21 × 760 × 0.79 mm Hg) or 3352 mm Hg and 12,608 mm Hg respectively. This is more than enough to cause oxygen toxicity and nitrogen narcosis.
II. The partial pressure of oxygen in the oxygen –helium mixture is (21 × 760 × 0.01 mm Hg) or 160 mm Hg. This is the same partial pressure of oxygen of inspired oxygen at the surface obviating the problem of oxygen toxicity.

(c) Helium is commonly used with oxygen (heliox) to dilute the oxygen in order to minimise the danger of oxygen toxicity.

Its other advantages are

I.   Lesser density as compared to nitrogen. Gas density is related to absolute pressure. As depth increases, the density of the breathing gas increases and becomes heavier per unit volume. Nitrogen with a higher density at depth will require a greater effort to breathe and restricts a diver's ability to ventilate the lungs effectively. A lower density gas like helium will alleviate this problem.

II. Heliox's low density produces a lower Reynolds number (<2000) resulting in laminar flow. Laminar flow generates less resistance when compared to turbulent flow and as a result the work of breathing is less when compared to breathing oxygen in compressed air at depth.
III. Helium molecules are relatively insoluble in lipids, minimising bubble liberation on decompression.

A point to note: Because sound travels faster in heliox than in air, it increases the pitch of the divers' speech, thereby making him speak like Donald duck.

8.1.8 Theme: Henry's Law, Boyle's Law and the 'Bends'

A deep sea scuba diver working at a depth of 100 m rapidly ascended to the surface due to a malfunctioning breathing apparatus. At the surface he complained of joint pains, blurring vision and dyspnoea. He is noted to be disorientated.

Question:

(a) What is Henry's Law and Boyle's Law and its relevance to respiratory physiology in diving?
(b) What is the pressure of inspired oxygen and nitrogen at sea level and at a depth of 100 m when breathing compressed air?
(c) What is the cause of his symptoms upon rapid ascending?
(d) How is this problem treated?

Answer:

(a) Henry's law states that the solubility of a gas in a liquid is proportional to the pressure of the gas over the solution. This means that as the ambient pressure increases the amount of gas dissolved in the blood and then subsequently in the tissues increases. In diving this is important in the genesis of nitrogen narcosis, oxygen toxicity and decompression sickness.

Boyle's law states that at constant temperature, the absolute pressure and the volume of a gas are inversely proportional. This can be restated in another manner, that is, the product of absolute pressure and volume is always constant.

Therefore as the diver descends, the gases in his body become smaller and as he ascends, they become bigger. This can give rise to problems known as the 'bends' and gas embolism.

(b) Dalton's Law states that the total pressure of a gas is equal to the sum of pressures of its individual components. Air contains 21 % oxygen and 78 % nitrogen.

Partial pressure of oxygen and nitrogen at sea level is

$$\text{Oxygen} : 0.21 \times 760 = 160 \, \text{mm Hg}$$
$$\text{Nitrogen} : 0.78 \times 760 = 593 \, \text{mm Hg}$$

Partial pressure of oxygen and nitrogen at a depth of 100 m is

Oxygen : $0.21 \times 760 \times 11$ atmosphere $= 1760\,mm\,Hg$

Nitrogen : $0.78 \times 760 \times 11$ atmosphere $= 6521\,mm\,Hg$

(c) The increased pressure of each gas component at depth will cause more of each gas to be dissolved into the blood and the body tissues (Henry's Law). This is exacerbated by the effect of Boyle's Law which decreases the volume of the gas molecules at depth.

Nitrogen is an inert gas and does not participate in the body's metabolism and as a result the amount of inhaled and exhaled nitrogen remains the same. This is unlike oxygen and carbon dioxide where the amount inhaled and exhaled are different, less oxygen and more carbon dioxide are exhaled. Oxygen does not contribute much to decompression illness as the excess oxygen is consumed by the body. Carbon dioxide is exhaled and also play no role in decompression illness.

The nitrogen that is inhaled and not metabolised will dissolve and saturate the blood and tissues. This is further aid by the relative high fat solubility of nitrogen.

No clinical sequelae occurs as the nitrogen remains in the dissolved phase under the high pressure environment at depth (Henry's Law).

However in an emergency rapid ascend, the ambient pressure will rapidly decrease causing large amount of nitrogen to come out of solution from the blood and the supersaturated tissues in the form of bubbles. Because nitrogen is highly soluble, a large volume of gas may be involved and the bubbles can affect any organ, such as the joints, the lungs, the heart and the brain. To exacerbate this problem, the bubbles also increase in size as the diver ascends (Boyle's Law). These bubbles are trapped in the tissues and blood stream in a process called 'embolism'. Life-threatening bubbles may block the pulmonary capillaries causing thoracic pain called chokes, dyspnoea and pulmonary oedema. CNS symptoms if the cranial vessels are blocked include paralysis, epileptic convulsions and unconsciousness. This is fatal if occurred underwater.

This phenomenon is known as the' Bends', decompression sickness (DCS) or Caisson disease and is primarily due to nitrogen bubbles. Apart from diving this phenomenon is also seen when a airplane at high altitude or a space craft suffers a sudden decompression

(d) The only effective treatment for decompression sickness is recompression in a hyperbaric chamber. All complications of decompression sickness are potentially reversible. Recompression returns the nitrogen to the dissolved state and decreases the size of the bubbles (Boyle's and Henry's Law). It allows a slower and controlled removal of excess nitrogen from the tissues.

Bibliography

Costanzo LS (2010) Physiology, 4th edn. Elsevier, Amsterdam

Davies A, Moores C (2010) The respiratory system: basic science and clinical conditions, 2nd edn. Elsevier, Amsterdam

Grippi MA (1995) Pulmonary pathophysiology. Lippincott, New York

Hall JE (2011) Guyton and hall textbook of medical physiology, 12th edn. Elsevier, Amsterdam

Levitzky MG (2003) Pulmonary physiology, 6th edn. McGraw Hill, New York

Marieb EN, Hoehn K (2014) Anatomy and physiology, 5th edn. Pearson, New York

Ming CH (2009) Respiratory physiology, figure-based instruction. Cengage Learning, Boston

Ming CH (2013) Conceptual learning in physiology. Pearson, New York

Raff H, Levitzky M (2011) Medical physiology, a systems approach. McGraw-Hill, New York

Sherwood L (2010) Human physiology, from cells to systems, 7th edn. Cengage Learning, Boston

Silverthorn DU (2012) Human physiology: an integrated approach, 6th edn. Pearson, New York

West JB (2008) Pulmonary pathophysiology, the essentials, 7th edn. Lippincott, New York

Widmaier EP, Raff H, Strang KT (2006) Vander's human physiology, 10th edn. McGraw-Hill, New York

© The Author(s) 2015
K.K. Mah and H.M. Cheng, *Learning and Teaching Tools for Basic and Clinical Respiratory Physiology*, SpringerBriefs in Physiology, DOI 10.1007/978-3-319-20526-7

Index

© The Author(s) 2015
K.K. Mah and H.M. Cheng, *Learning and Teaching Tools for Basic and Clinical Respiratory Physiology*, SpringerBriefs in Physiology, DOI 10.1007/978-3-319-20526-7

Printed in the United States
By Bookmasters